T0245316

Haematology Case Studies with Blood Cell Morphology and Pathophysiology

Haematology Case Studies with Blood Cell Morphology and Pathophysiology

Indu Singh, PhD
Griffith University, QLD, Australia

Alison Weston, BSc, Dip Ed(Adult)
Griffith University, QLD, Australia

Avinash Kundur, PhD
Griffith University, QLD, Australia

Gasim Dobie, PhD
RMIT University, VIC, Australia

ACADEMIC PRESS

An imprint of Elsevier

Academic Press is an imprint of Elsevier
125 London Wall, London EC2Y 5AS, United Kingdom
525 B Street, Suite 1800, San Diego, CA 92101-4495, United States
50 Hampshire Street, 5th Floor, Cambridge, MA 02139, United States
The Boulevard, Langford Lane, Kidlington, Oxford OX5 1GB, United Kingdom

Library of Congress Cataloging-in-Publication Data
A catalog record for this book is available from the Library of Congress

British Library Cataloguing-in-Publication Data
A catalogue record for this book is available from the British Library

ISBN 978-0-12-811911-2

For information on all Academic Press publications
visit our website at https://www.elsevier.com/books-and-journals

Working together
to grow libraries in
developing countries

www.elsevier.com • www.bookaid.org

Publisher: Mica Haley
Acquisition Editor: Tari K. Broderick
Editorial Project Manager Intern: Gabriela D. Capille
Production Project Manager: Punithavathy Govindaradjane
Cover Designer: Greg Harris

Typeset by SPi Global, India

Contents

Foreword

The aim of this book is to present the haematology morphology of a Romanowsky-stained peripheral blood smear which may be used in conjunction with a text book to assist with the diagnosis of blood cell diseases.

"Case Studies with Cell Morphology and Pathophysiology" is divided into very useful and well-thought-out sections with excellent photographic examples of the morphology in disease and an array of the "typical" haematology results and pathophysiology for each condition. Even with major advances in haematology automation and molecular diagnostics, examination of a properly made and stained peripheral smear is still considered to be the "old standard" in diagnosing blood cell disorders. Sadly, this appears to be a diminishing skill for many scientists.

This reference is a valuable adjunct to the libraries of those beginning students in haematology morphology, experienced morphologists, to the multi-disciplinary scientists wishing to enhance their morphology skills and to clinicians alike.

Robert J. Horwood
B. Applied Science (MLS); Dip. Ed.; Cert IV TESOL; MAIMS
Haematology Scientist (Retired)

Preface

This book is written for students of laboratory medicine and haematology both at undergraduate and postgraduate level as well as for people who work in the clinical or hospital diagnostic laboratories and want to increase their knowledge in haematology, particularly blood cell morphology and associated pathophysiology of haematological disorders. This book will provide basic information about how to recognize and diagnose some haematological conditions that are frequently observed in the laboratories. The population of technicians and scientists who will benefit most from this book include those working in core laboratories including biochemistry or blood bank and are rotated around various disciplines as part of shift and weekend work. Moreover, it can be used as a reference book by technicians, scientists, and haematologists alike because it includes information relevant to every level of expertise in diagnosing haematological disorders. Most haematology books or atlases available in the market are either very in-depth and long or very basic with blood cell images with minimal explanation. We have compiled in this book specific case studies with general and specific information about various haematological disorders with Full Blood Examination (FBE or CBC), blood film images, pathophysiology of the conditions, and further confirmatory evaluation with their expected results for final diagnosis. This is what makes our book unique but simple. This book is the result of demand from industry for all in one simple reference book for teaching morphological skills to new trainees and upgrading the skills of senior workers starting rotation between various disciplines of laboratory as increasingly seen in the current economic situation.

Acknowledgements

This book has taken 7 years to complete and we are gratful for the input and encouragement of everyone along the way. A special mention must be made to RMIT for the use of their blood film collection and Ms. Janine Myers from Royal Melbourne Hospital. Other people to be mentioned are Ms. Pamela Taylor whose attention to detail helped with editing the text and the images and Ms. Quirine O'Loughlin who took time out of her busy schedule to review the completed text. Many thanks also to all of our colleagues for their suggestions and discussions about the clinical conditions included in the book.

Chapter 1

Introduction

THE IMPORTANCE OF THE BLOOD FILM IN THE DIAGNOSIS OF HAEMATOLOGICAL DISORDERS

Diagnostic information can be obtained from the blood smear examination if done by an experienced laboratory scientist or haematologist. Aside from the fact that it allows certain organisms to be directly visualized, peripheral blood film morphology gives helpful information on the aetiology, severity, host susceptibility, and systemic impact. It assists us to see the typical footprints left by different infections on blood cell morphology and this gives cytological clues in diagnosis such as in the case of Dohle bodies in haemophagocytosis. Physicians usually request a blood smear following perceived clinical features or irregularity of a previously done full blood count. Several factors determine whether a smear will be reviewed or not. These include the sex and age of the patient, whether such a request is an initial or subsequent one and the degree of clinical improvement from the previous result. It is possible to make a definitive diagnosis from the blood smear such as in the case of haemolytic anaemia, myelodysplastic syndrome (MDS), leukaemia, and lymphoma [1]. The morphological features of the red blood cell (RBC) and white blood cell (WBC) are important in the diagnosis of haemolytic anaemia, leukaemia, and lymphoma. The aetiology and confirmation of such diseases is then possible. Differential diagnosis can be considered which may make further investigation necessary. This booklet identifies the major roles of the blood smear in the diagnosis and differential diagnoses of anaemia, thrombocytopenia, as well as the determination and characterization of lymphoma and leukaemia. Furthermore, it highlights the further tests that should be done to confirm the diagnosis and the pathophysiology behind the condition.

MICROCYTIC ANAEMIA, HAEMOGLOBINOPATHIES, AND BLOOD FILM MORPHOLOGY

Most cases of microcytic anaemia are diagnosed by a combination of red cell indices, inflammatory markers, serum ferritin level, and clinical suspicion. However, the presence of Pappenheimer bodies and dimorphism of RBCs are helpful in the case of sideroblastic anaemia, basophilic stippling in lead poisoning, and some types of thalassemia. In the case of iron deficiency anaemia,

Haematology Case Studies with Blood Cell Morphology and Pathophysiology. http://dx.doi.org/10.1016/B978-0-12-811911-2.00001-5

the presence of elongated cells is the most diagnostic feature. The blood smear is also important in the diagnosis of sickle cell trait as compound heterozygosity of haemoglobin S and haemoglobin C may have a normal haemoglobin level and thus may be confused with sickle cell trait. The blood smear of a compound heterozygote usually shows irregular contracted cells, target cells, and boat-shaped cells with few classic sickle cells. So, the blood smear combined with sickle solubility test permits an accurate diagnosis [1].

BLOOD FILM MORPHOLOGY AND DIAGNOSIS OF MACROCYTIC ANAEMIA

Hypersegmented neutrophils, macrocytes, as well as macroovalocytes are seen in macrocytic anaemia induced by Vitamin B12 or folic acid deficiency. In the severe form of this, one may also see red cell fragments and tear drop poikilocytes in the blood film. The blood smear is important here in that it provides provisional speedy diagnosis and allows for commencement of therapy while awaiting results. It is also useful in that it eliminates problems of false negativity of patients with remarkable B12 deficiencies that however have normal assay result. This occurs because a lot of the B12 being assayed for is bound to haptocorrin while the effective B12 bound to transcobalamin less contributes to the total B12 assay. Hepatic disease and alcohol intake usually give rise to macrocytic anaemia. In this case, the blood smear shows round (not oval) macrocytes without any hypersegmented neutrophils. However, stomatocytes and targets may occasionally be present [1]. In older individuals, macrocytosis may be caused by myelodysplastic syndromes with the blood smear often revealing hypolobulated and hypogranular neutrophils. Other morphological features include Pappenheimer bodies, blast cells, and giant platelets. Sometimes hypochromic microcytes are present which makes the smear dimorphic. In macrocytosis brought about by blood loss or recent haemolysis, RBCs usually have polychromasia due to reactionary reticulocytosis. Significant poikilocytosis is seen sometimes in macrocytic anaemia [1].

PERIPHERAL BLOOD FILM AND DIAGNOSIS OF HAEMOLYTIC ANAEMIA

The shape of the RBC is important in the diagnosis of haemolytic anaemia. Certain types of haemolytic anaemia yield distinctive blood smears which can be enough for diagnosis. These include hereditary spherocytosis, hereditary pyropoikilocytosis, as well as South East Asia ovalocytosis. In hereditary elliptocytosis, the blood smear shows several elliptocytes as well as a smaller number of ovalocytes. In hereditary pyropoikilocytosis, one can see numerous poikilocytes including ovalocytes, elliptocytes, and fragmented cells. South East Asia ovalocytosis reveals poikilocytosis with macroovalocytes seen after staining specimens with May-Grunwald Giemsa stain [1]. Hyperchromic small cells are seen in conditions such as microangiopathic haemolytic anaemia burns as well as spherocytic haemolytic anaemia. The diagnosis of conditions such as disseminated cancer, haemolytic uremic syndrome (HUS), pregnancy-induced hypertension, and thrombotic thrombocytopenic purpura (TTP) is aided with the detection of a microangiopathic haemolytic anaemia. Furthermore, the blood film becomes important in diagnosing acute haemolysis brought about by oxidative damage. Morphologically, one would be able to see bite cells or keratocytes. Blister cells as well as irregularly contracted cells are also seen. Oxidative stress leading to haemolysis is seen in anaemia of G6PD deficiency. In the case of G6PD deficiency, G6PD assay can be normal as seen in acute haemolysis in G6PD-deficient individuals who usually are men of African-American origin or those females who are carriers. In these two categories, a repeat assay is indicated after the period of acute haemolytic episode. The red cell in G6PD deficiency is also polychromatic and irregularly contracted. For hereditary spherocytosis, there are numerous, hyperchromic cells with a regular spherocyte outline. The blood film in diagnosing G6PD becomes more important even when G6PD assay is normal [1].

BLOOD FILM MORPHOLOGY, THROMBOCYTOPENIA, AND THROMBOCYTOSIS

Blood film examination is also important in individuals with thrombocytosis to check for the possibility of a myeloproliferative disorder change such as giant platelets or basophilia. The basophilia is not reliably detected by an automated counter and therefore requires clarification by examination of a blood smear. Falsely low platelet counts can be seen with platelet clumping and satellitism as well as in small clots, whereas falsely high platelet counts may be the result of red cell fragments, leukaemic cell fragments, and fungi being counted as platelets. The presence of fibrin strands suggests that thrombocytopenia may be factitious [1].

BLOOD FILM MORPHOLOGY, LEUKAEMIA, LYMPHOMA, AND BONE MARROW FAILURE

In the case of unexplained monocytosis, leucocytosis, and lymphocytosis, it is important to check the blood film. Pancytopenia in which all the cell lines are low is also an indication for blood film analysis. This may be due to an acute leukaemia, bone marrow infiltration by malignant cells, hairy cell leukaemia, or aplastic anaemia. The blood smear helps to indicate which further tests should be carried out. This also provides a morphologic basis in the absence of which sophisticated investigations and immunophenotyping cannot be interpreted. In acute promyelocytic leukaemia and Burkett's lymphoma, blood smears can help make rapid diagnosis and specific treatment can thus be initiated rarely without further flow results. In Burkett's lymphoma, the blood smear shows basophilic vacuolation in lymphocytic cells while Auer rods are seen in acute promyelocytic leukaemia [1].

BRIEF ABOUT LEUKAEMIA

Leukaemia is a disease that affects blood-forming cells in the body. It is a cancerous condition characterized by an abundance of abnormal blood cells including leucocytes, erythrocytes, and thrombocytes blood cells in the body. Leukaemia begins in the bone marrow and spreads to other parts of the body. Both children and adults can develop leukaemia. The other condition that can affect the white blood cells is lymphoma which originates from the lymphatic system [2]. The direct causes related to the development of the malignancy are unknown; however, there are some conditions that may cause leukaemia. Environmental toxins can induce genetic changes, leading to the malignancy; exposure to radiation is also known to lead to malignancy. Viruses such as Epstein–Barr virus may play a pathogenetic role in causing a lymphoid malignancy. Additionally, it is considered that some alkylating agents and other chemotherapy that are used to treat various forms of cancer can cause hematologic malignancy as they induce DNA damage in hematopoietic cells. Generally, genetic changes or mutations could be the reason behind the leukaemia regardless of the causes of these mutations. The classification and diagnosis of leukaemia under the FAB system were based largely on morphologic characteristics and cytochemistry. However, there is a movement towards another classification to be more precise because of recurring chromosomal and genetic lesions that were found in many patients. These lesions are related to disruption of oncogenes which are genes that cause dominant-acting cancer mutation; tumour suppressor genes which code for proteins that help cells resist malignant transformation; and other regulatory elements that control apoptosis (programmed cell death), proliferation, maturation, and other vital cell functions. In 2001 the WHO (world health organisation) included or added the chromosomal translocation to the old criteria. The study of chromosomal translocation in hematopoietic malignancies has taught us how a single mutation or series of mutations may result in malignant transformation by disrupting the molecular machinery of the cell. There are many chromosomal translocations occurring in leucocyte malignancies. For example, t(9:22) in chronic myelogenous leukaemia which was the first genetic lesion found in human cancer, whereas the second one was t(8:14) in Burkett's lymphoma [2].

EXPLANATIONS FOR SOME TERMS USED IN THIS BOOKLET

Provisional diagnosis is a temporary and most likely diagnosis that needs to be confirmed by further appropriate tests.

Differential diagnosis means that there is a possibility that the diagnosis could be something else such as other disorders which may share the same symptoms of the provisional diagnosis and they should be ruled out by further tests to reach the actual diagnosis.

Further tests are a number of tests that should be done to confirm the provisional diagnosis and exclude the differential diagnosis.

Pathophysiology is the study of the changes of normal mechanical, physical, and biochemical functions, either caused by a disease or resulting from an abnormal syndrome.

CYTOGENETICS

Cytogenetics is the study of the structure and the inheritance of the chromosomes (r408). Normally, there are 46 chromosomes found in each cell [2]. Each chromosome is composed of a short arm (p) and a long arm (q) separated by a region known as the centromere. The classification of chromosome disorders can be done as structural or numerical and involve the gain, loss, or the rearrangement of the chromosomes [2].

Normal chromosome number (46) = diploid, 23 pairs and there are two arms: short (p) and long (q).

(t) Translocation or exchange between two or more chromosomes.

For example; t(12:21) (p13:q22) → (t) represents a translocation and the two numbers in the brackets represent the number of the chromosome, (p) represents the short arm and (q) represents the long arm.

(Hyperdiploidy) → gain of a chromosome—trisomy
(Hypodiploidy) → loss of a chromosome—monosomy
(del) → deletion—loss of part of a chromosome
(inv) → inversion—rearrangement within an individual chromosome

CYTOCHEMISTRY

Cytochemistry is the study of chemical elements found in the cells. These elements can be enzymes, lipids, or glycogen. It can be used in differentiating hematopoietic malignancies, especially acute and chronic leukaemias [2]. There are many stains that are used in cytochemistry tests.

MYELOPEROXIDASE (MPO)

MPO is a peroxidase enzyme present in primary granules of neutrophils, eosinophils, and to a certain extent of monocytes. Lymphocytes do not exhibit MPO activity [2]. So, immunohistochemical staining for myeloperoxidase used to be administered in the diagnosis of acute myeloid leukaemia to demonstrate that the leukaemic cells were derived from the myeloid lineage. It differentiates between acute myeloid leukaemia and acute lymphoblastic leukaemia [2].

SUDAN BLACK B (SBB)

SBB is a lysochrome diazo dye used for staining of lipids, such as neutral triglycerides, sterols, and phospholipids [2].

In differentiating haematological disorders Sudan black will stain myeloblasts but not lymphoblasts, because these lipids are found in the neutrophil granules and in the monocyte lysosomal granules [2]. So, it differentiates acute myeloid leukaemia (AML) from acute lymphoblastic leukaemia (ALL).

CHLOROACETATE ESTERASE

The aim of this stain is to demonstrate the presence of granulocytes. The activity of chloroacetate esterase is found in the neutrophils and their precursors [3]. It differentiates between acute myeloid leukaemia and acute lymphoblastic leukaemia.

PERIODIC ACID SCHIFF (PAS)

PAS staining is mainly used for staining structures containing a high proportion of carbohydrate including glycogen, which is often found in hematopoietic cells [3]. This stain is positive in the majority of acute lymphoblastic leukaemias and negative in the majority of myeloid and monoblastic leukaemias [3]. Thus it differentiates ALL from AML.

TERMINAL DEOXYNUCLEOTIDYL TRANSFERASE STAIN (TdT)

The TdT stain differentiates between acute lymphoblastic leukaemia and acute myelogenous leukaemia as it is positive in ALL [2].

NEUTROPHIL ALKALINE PHOSPHATASE (NAP)

The NAP score is used to differentiate between chronic myelogenous leukaemia (CML) and other myeloproliferative neoplasms (MPN) as it is low in CML and normal or high in other MPN, and between CML and leukemoid reaction as it increases in leukemoid reactions [3].

TARTRATE-RESISTANT ACID PHOSPHATASE STAIN (TRAP)

The TRAP stain is primarily used to identify the cell of hairy cell leukaemia cells [2].

ACID PHOSPHATASE

Acid phosphatase is usually positive in T-cell leukaemia [3].

ALPHA NAPHTHYL ACETATE ESTERASE (ANAE)

Alpha Naphthyl Acetate Esterase confirms monocytic differentiation [3].

IMMUNOPHENOTYPING

Immunophenotyping is the analysis of heterogeneous populations of cells for the purpose of identifying the presence and proportions of the various populations of interest. Antibodies are used to identify cells by detecting specific antigens expressed by these cells, which are known as markers. These markers are usually functional membrane proteins involved in cell communication, adhesion, or metabolism. Immunophenotyping using flow cytometry has become the method of choice in identifying and sorting cells within complex populations, for example, the analysis of immune cells in a blood sample. Applications of this technology are used both in basic research and clinical laboratories. Cell markers are a very useful way to identify a specific cell population. However, they will often be expressed on more than one cell type. Therefore flow cytometry staining strategies have led to methods for immunophenotyping cells with two or more antibodies simultaneously. By evaluating the unique repertoire of cell markers using several antibodies together, each coupled with a different fluorochrome, a given cell population can be identified and quantified. Many immunological cell markers are CD markers and these are commonly used for detection in flow cytometry of specific immune cell populations and subpopulations. CD (Cluster of Differentiation) markers are a group of special molecules on the surface of the cells in our body. There are several types of CD molecules. All cells in our body have one or more of them, but they are most useful for classifying WBC (white blood cells) malignancies as every linage expresses specific antigens. For example, the myeloblast is characterized by expression of immature cell markers CD34, CD38, HLA-DR, and stem cell factor receptor CD117. Pan-myeloid markers, CD13 and CD33, present on all myeloid progeny are first expressed at this stage. In monocytic lineage, bright expression of CD64 and HLA-DR antigens persists throughout monocytic maturation. In addition, Glycophorin A is present on reticulocytes and erythrocytes whereas CD41 and CD61 appear as the first markers of megakaryocytic differentiation. In regard to lymphoid lineage, the lymphoid progenitors express CD34, terminal deoxynucleotidyl transferase (TdT), and HLA-DR as the blast cells. CD19, CD22, and CD79 are expressed in the B-lineage whereas CD2, CD3, and CD7 appear in the T-cell lineage [2].

List of the Most Important Haematological Conditions Discussed in This Booklet

1. Mild IDA
2. Marked IDA
3. A-Thal minor
4. B-Thal minor
5. B-Thal major
6. Haemoglobin H disease
7. Hb-E disease
8. Sideroblastic anaemia
9. Hb S disease and sickle cell with Thal
10. Anaemia of chronic disease
11. Aplastic anaemia
12. Post splenectomy change
13. Macrocytosis
14. Megaloblastic anaemia
15. Liver disease
16. G6PD deficiency
17. PK deficiency
18. HS in crisis
19. HE
20. H. stomatocytosis
21. Melanesian ovalocytosis
22. Hereditary pyropoikilocytosis
23. PNH
24. DIHA
25. Warm-AIHA
26. Cold-AIHA
27. HDNB-Rh
28. AML W/O maturation
29. AML with maturation
30. APML
31. AMML
32. Acute monoblastic leukaemia
33. Acute monocytic leukaemia
34. Acute erythroid leukaemia
35. Acute megakaryocytic leukaemia
36. ALL
37. CML
38. PV
39. ET
40. MF
41. CMML
42. MDS-RARS
43. MDS-RAEB
44. CLL
45. CLL+AIHA
46. PLL
47. HCL
48. Burkitt's lymphoma
49. Chronic mantle cell lymphoma (MCL)
50. Plasma cell leukaemia (PCL)
51. Plasma cell myeloma
52. DIC
53. HUS
54. TTP
55. HELLP
56. IM
57. Marked left shift with marked toxic granulation
58. Malaria (*Plasmodium vivax*)
59. Malaria (*Plasmodium falciparum*)
60. Malaria combined (*Plasmodium falciparum+Plasmodium vivax*)
61. Malaria (*Plasmodium ovale*)
62. Malaria (*Plasmodium malariae*)

Chapter 2

Microcytic Disorders

Case 1

Clinical presentations

- Fatigue
- Pallor
- Weakness
- Hair loss

FBE RESULTS

CBC (Complete Blood Count) Parameters

Parameters	Results	Units
Hb	100	g/L
RCC	4.00	$\times 10^{12}\,L^{-1}$
PCV	0.30	
MCV	75	Fl
MCH	25.0	Pg
MCHC	333	g/L
RDW	19.0	%
Retics	N/A	%
WBC	6.0	$\times 10^{9}\,L^{-1}$
Plt count	200	$\times 10^{9}\,L^{-1}$

Haematology Case Studies with Blood Cell Morphology and Pathophysiology. http://dx.doi.org/10.1016/B978-0-12-811911-2.00002-7

Differential White Cell Count

Cell Type	Percentages (%)	Absolute Count
Neutrophils	64	$3.84 \times 10^9 \, L^{-1}$
Bands	0	$0 \times 10^9 \, L^{-1}$
Lymphocytes	21	$1.26 \times 10^9 \, L^{-1}$
Monocytes	11	$0.66 \times 10^9 \, L^{-1}$
Eosinophils	3	$0.18 \times 10^9 \, L^{-1}$
Basophils	1	$0.06 \times 10^9 \, L^{-1}$
Metamyelocytes	0	
Myelocytes	0	
Promyelocytes	0	
Blasts	0	
Prolymphocytes	0	
nRBC/100WBC	0	

N/A, not available.

RBC MORPHOLOGY

Moderate anisocytosis, mild microcytes, moderate hypochromic, moderate elongated, mild irregular contracted cells, and mild ovalocytes.

WBC MORPHOLOGY

Normal in number and morphology.

PLATELET MORPHOLOGY

Normal in number and morphology.

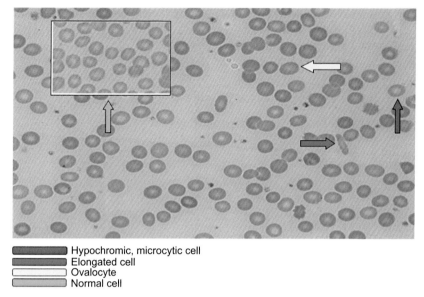

Hypochromic, microcytic cell
Elongated cell
Ovalocyte
Normal cell

Blood film image shows the most diagnostic features.

PROVISIONAL DIAGNOSIS

Based on clinical presentations, FBE results, and morphology, this case is highly suggestive of mild iron deficiency anaemia and further tests are suggested.

DIFFERENTIAL DIAGNOSES

Normal
Thalassemias/other haemoglobinopathies
Congenital sideroblastic anaemia
Anaemia of chronic disease
Combination of IDA with other condition

FURTHER TESTS AND EXPECTED RESULTS

Confirmation of Provisional Diagnosis

Iron studies

- Serum iron → low (normal range 10–30 μmol/L)
- Total iron binding capacity (TIBC) → High (normal range 40–70 μmol/L)
- % saturation → < 10%

- Ferritin → low (normal range 50–150 µg/dL) (could be high if there is coexistent inflammatory disorder).
- Soluble transferrin receptor → increased
- Bone marrow iron → absent/deficient (present in sideroblastic anaemia)

Exclusion of Differential Diagnoses

HPLC and haemoglobin electrophoresis →

Hb A—98%
Hb F—less than 1%
Hb A2—within normal range (2%–3%)

FINAL DIAGNOSIS

Mild iron deficiency anaemia is the final diagnosis for this case based on further tests that confirm the provisional diagnosis and exclude all the differential diagnoses.

Case 2

Clinical presentations

- General features of anaemia, including dyspnoea on effort, weakness, and dizziness.
- Angular stomatitis and atrophic glossitis.
- Brittle spoon-shaped nail (koilonychias).
- Brittle sparse hair.

FBE RESULTS

CBC (Complete Blood Count) Parameters

Parameters	Results	Units
Hb	60	g/L
RCC	3.82	$\times 10^{12}\,L^{-1}$
PCV	0.21	
MCV	55	Fl
MCH	15.7	Pg
MCHC	284	g/L
RDW	19.1	%
Retics	N/A	%
WBC	7.9	$\times 10^{9}\,L^{-1}$
Plt count	569	$\times 10^{9}\,L^{-1}$

Differential White Cell Count

Cell Type	Percentages (%)	Absolute Count
Neutrophils	61	$4.82 \times 10^{9}\,L^{-1}$
Bands	0	$0 \times 10^{9}\,L^{-1}$
Lymphocytes	25	$1.97 \times 10^{9}\,L^{-1}$
Monocytes	11	$0.87 \times 10^{9}\,L^{-1}$
Eosinophils	3	$0.24 \times 10^{9}\,L^{-1}$
Basophils	0	$0 \times 10^{9}\,L^{-1}$
Metamyelocytes	0	

Continued

Differential White Cell Count—cont'd

Cell Type	Percentages (%)	Absolute Count
Myelocytes	0	
Promyelocytes	0	
Blasts	0	
Prolymphocytes	0	
nRBC/100WBC	0	

RBC MORPHOLOGY

Moderate anisocytosis, marked microcytosis, moderate hypochromia, moderate elongated cells, and occasional target cells.

WBC MORPHOLOGY

Normal in number and morphology.

PLATELET MORPHOLOGY

Mild thrombocytosis with normal morphology.

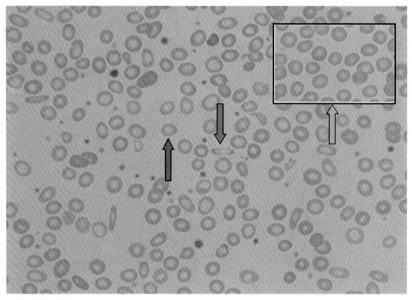

☐ Hypochromic, microcytic cell
☐ Elongated cell
☐ Normal cells

Blood film image shows the most diagnostic features.

PROVISIONAL DIAGNOSIS

Based on clinical presentations, FBE results, and morphology, this case is highly suggestive of marked iron deficiency anaemia and further tests are suggested.

DIFFERENTIAL DIAGNOSES

> Thalassemias/other haemoglobinopathies
> Congenital sideroblastic anaemia
> Anaemia of chronic disease
> Combination of IDA with other condition

FURTHER TESTS AND EXPECTED RESULTS

Confirmation of Provisional Diagnosis

Iron studies

- Serum iron → low (normal range 10–30 μmol/L)
- Total iron binding capacity (TIBC) → High (normal range 40–70 μmol/L)
- % saturation → < 10%
- Ferritin → low (normal range 50–150 μg/dL) (could be high if there is coexistent inflammatory disorder).
- Soluble transferrin receptor → increased
- Bone marrow iron → absent/deficient (present in sideroblastic anaemia)

Exclusion of Differential Diagnoses

HPLC and haemoglobin electrophoresis →

> Hb A—98%
> Hb F—less than 1%
> Hb A2—within normal range (2%–3%)

FINAL DIAGNOSIS

Marked iron deficiency anaemia is the final diagnosis for this case based on further tests that confirm the provisional diagnosis and exclude all the differential diagnoses.

PATHOPHYSIOLOGY OF IDA

Iron deficiency anaemia develops slowly, progressing through stages that physiologically blend one into the other but are a useful delineation for understanding disease progression. Iron is distributed among three compartments: (1) the storage compartment, principally as ferritin in the bone marrow macrophages and liver cells; (2) the transport compartment of serum transferrin; and (3) the functional compartment of haemoglobin (Hb), myoglobin, and cytochromes. Hb and intracellular ferritin constitute nearly 95% of the total distribution of iron [2].

Iron absorption: Iron is absorbed in the duodenum and upper jejunum. Absorption of iron is determined by the type of iron molecule and by what other substances are ingested. Iron absorption is best when food contains haeme iron. Dietary nonhaeme iron must be reduced to the ferrous state and released from food binders by gastric secretions. Nonhaeme iron absorption is reduced by other food items and certain antibiotics.

Iron transport and usage: Iron from intestinal mucosal cells is transferred to transferrin, an iron-transport protein synthesized in the liver; transferrin can transport iron from cells (intestinal, macrophages) to specific receptors on erythroblasts, placental cells, and liver cells. For haeme synthesis, transferrin transports iron to the erythroblast mitochondria, which insert the iron into protoporphyrin for it to become haeme. Transferrin is extruded for reutilization. Synthesis of transferrin increases with iron deficiency but decreases with any type of chronic disease.

Iron storage and recycling: Iron not used for erythropoiesis is transferred by transferrin, an iron transporting protein, to the storage pool which has two forms, ferritin and haemosiderin. The most important form is ferritin which is a soluble and active storage fraction located in the liver (in hepatocytes), bone marrow, and spleen (in macrophages); in RBCs; and in serum. Iron stored in ferritin is readily available for any body requirement. Circulating (serum) ferritin level parallels the size of the body stores. The second storage pool of iron is in haemosiderin, which is relatively insoluble and is stored primarily in the liver (in Kupffer cells) and in the marrow (in macrophages).

Because iron absorption is so limited, the body recycles and conserves iron. Transferrin grasps and recycles available iron from ageing RBCs undergoing phagocytosis by mononuclear phagocytes. This mechanism provides about 97% of the daily iron needed (about 25 mg of iron). With ageing, iron stores tend to increase because iron elimination is slow [4].

For a period of time as iron intake lags behind loss, essentially normal iron status continues. Absorption of iron through the intestine is accelerated in an attempt to meet the relative increased demand for iron, but this is not apparent in laboratory tests or patient symptoms. The individuals appear healthy. As the negative iron balance continues, however, storage of iron depletion develops.

Stage 1. Stage 1 of iron deficiency is characterized by a progressive loss of storage iron. The body's reserve of iron is sufficient to maintain the transport and functional compartment through this phase, so red blood cell (RBC) development is normal. There is no evidence of iron deficiency in the peripheral blood picture and the patient experiences no symptoms of anaemia. If ferritin levels are measured, however, they are low, indicating the decline in stored iron, which also could be detected in an iron stain of the marrow. Without evidence of anaemia, however, neither of these tests would be performed, and individuals appear healthy.

Stage 2. Stage 2 of iron deficiency is defined by the exhaustion of the storage pool of iron. For a time RBC production continues as normal, relying on the iron available in the iron transport compartment. Anaemia, as measured relative to the reference range of Hb, is still not evident, although an individual's Hb may begin dropping. Other iron-dependent tissues, such as muscles may begin to be affected, although the symptoms may be nonspecific. If measured, ferritin levels and serum iron are still low, whereas total iron-binding capacity (TIBC) increases. Free erythrocyte protoporphyrin (FEP), the porphyrin into which iron is inserted into to form haeme, begins to accumulate. Transferrin receptors increase on the surface of iron-starved cells as they try to capture as much available iron as possible. They also are shed into the plasma, and levels increase measurably in stage 2. Prussian blue stain of the marrow in stage 2 would show essentially no stored iron and iron-deficient erythropoiesis would be evident. As in stage one, iron deficiency in stage 2 is subclinical, and testing is not likely to be undertaken.

Stage 3. Stage 3 of iron deficiency is frank anaemia. The Hb and haematocrit (Hct) are low relative to the reference ranges. Having thoroughly depleted storage iron and diminished transport iron, developing RBCs are unable to develop normally. The number of cell divisions per precursor increases because Hb accumulation in the developing cells is slowed, allowing more time for divisions. The result is at first smaller cells with adequate Hb concentration, although ultimately even these cannot be filled with Hb. These cells become microcytic and hypochromic. As would be expected, ferritin levels are exceedingly low. Other iron studies also are abnormal, and the FEP and transferrin receptors levels increase. In this phase, the patient experiences the nonspecific symptoms of anaemia, typically fatigue and weakness, especially with exertion. Pallor is evident in light-skinned individuals but also can be noted in the conjunctive, mucus membranes, or palmar creases of dark-skinned individuals [2].

Case 3

Clinical presentation

- No serious clinical symptoms

FBE RESULTS

CBC (Complete Blood Count) Parameters

Parameters	Results	Units
Hb	117	g/L
RCC	5.12	$\times 10^{12} L^{-1}$
PCV	0.40	
MCV	78	Fl
MCH	22.9	Pg
MCHC	292	g/L
RDW	14.8	%
Retics	N/A	%
WBC	7.0	$\times 10^{9} L^{-1}$
Plt count	200	$\times 10^{9} L^{-1}$

Differential White Cell Count

Cell Type	Percentages (%)	Absolute Count
Neutrophils	70	$4.9 \times 10^{9} L^{-1}$
Bands	4	$0.28 \times 10^{9} L^{-1}$
Lymphocytes	11	$0.77 \times 10^{9} L^{-1}$
Monocytes	13	$0.91 \times 10^{9} L^{-1}$
Eosinophils	1	$0.07 \times 10^{9} L^{-1}$
Basophils	1	$0.07 \times 10^{9} L^{-1}$
Metamyelocytes	0	

Differential White Cell Count—cont'd		
Cell Type	Percentages (%)	Absolute Count
Myelocytes	0	
Promyelocytes	0	
Blasts	0	
Prolymphocytes	0	
nRBC/100WBC	0	

RBC MORPHOLOGY

Moderate anisocytosis, moderate microcytes, moderate hypochromic, mild elongated cells, moderate target cells, mild rouleaux, and mild crenated cells.

WBC MORPHOLOGY

Normal in number and morphology.

PLATELET MORPHOLOGY

Normal in number and morphology.

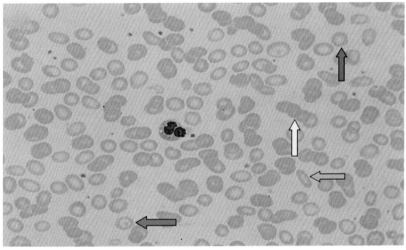

Hypochromic, microcytic cell
Target cell
Rouleaux
Elongated cell

Blood film image shows the most diagnostic features.

PROVISIONAL DIAGNOSIS

Based on clinical presentations, FBE results, and morphology, this case is highly suggestive of alpha thalassemia minor and further tests are suggested.

DIFFERENTIAL DIAGNOSES

Beta thalassemia minor
Iron deficiency anaemia
Haemoglobin constant spring

FURTHER TESTS AND EXPECTED RESULTS

Confirmation of Provisional Diagnosis

Haemoglobin electrophoresis and HPLC → normal except during the neonatal period. Haemoglobin A2 → normal/low
Family history → positive
Molecular tests → deletion of two α-globin genes

Exclusion of Differential Diagnoses

Iron profile

Serum iron → normal
TIBC → normal
Ferritin → normal

Exclude haemoglobin Constant Spring by Hb-electrophoresis and HPLC → should be normal.

FINAL DIAGNOSIS

Alpha thalassemia minor is the final diagnosis for this case based on further tests that confirm the provisional diagnosis and exclude all the differential diagnoses.

PATHOPHYSIOLOGY OF ALPHA THALASSEMIA

The absence of normal production of α-chains results in a relative excess of γ-globin chains in the foetus and newborn, and β-globin chains in children and adults. Further, the β-globin chains are capable of forming soluble tetramers (β-4, or HbH); yet this form of haemoglobin is unstable and tends to precipitate within the cell forming insoluble inclusions (Heinz bodies) that damage the red cell membrane. Furthermore, diminished haemoglobinization of individual red blood cells results in damage to erythrocyte precursors and ineffective erythropoiesis in the bone marrow, as well as hypochromia and microcytosis of circulating red blood cells.

Genes that regulate both synthesis and structure of different globins are organized into two separate clusters. The α-globin genes are encoded on chromosome 16 and the γ, δ, and β-globin genes are encoded on chromosome 11. Each individual normally carries a linked pair of α-globin genes, two from the paternal chromosome, and two from the maternal chromosome. Alpha thalassemia results when there is disturbance in production of α-globin from any or all four of the α-globin genes.

Normal haemoglobin biosynthesis requires an intact gene, silencers, enhancers, promoters, and locus control region (LCR) sequences. Several hundred mutations causing thalassemia have been described. These may affect any step in globin gene expression, transcription, pre-mRNA splicing, mRNA translation and stability, and post-translational assembly and stability of globin polypeptides.

The most common mechanism of aberrant α-globin production is due to deletions of either portions of the α-globin genes themselves or the genetic regulatory elements that control their expression. Regulatory elements may be located on the same chromosome (cis-acting elements) or on separate chromosomes (trans-acting elements).

Production of functional haemoglobin is also impaired in alpha thalassemia when point mutations, frame shift mutations, nonsense mutations, and chain termination mutations occur within or around the coding sequences of the α-globin gene cluster. These gene level mutations may in turn affect RNA splicing, initiation of mRNA translation, or result in the generation of unstable α-chain variants.

Mutations affecting transcription, pre-mRNA splicing, or canonical splice signals are rare causes of alpha thalassemia. Other forms of alpha thalassemia are caused by either premature or failed translation termination. More rare mutations have been found to cause thalassemia by interfering with the normal folding of otherwise normal globin peptide.

From the genetic standpoint, alpha thalassemia are extremely heterogenous: however, the phenotypic expression of alpha thalassemia may be described as in simplified clinical terms related to the number of α-globin genes affected:

> Alpha (0) thalassemia—More than 20 different genetic mutations that result in the functional deletion of both pair of α-globin genes have been identified. Individuals with this disorder are not able to produce any functional α-globin and thus are unable to make any functional haemoglobin A, F, or A2. This leads to the development of hydrops fetalis, also known as haemoglobin Bart, a condition that is incompatible with extra uterine life.
>
> Alpha (+) thalassemia—There are more than 15 different genetic mutations that result in decreased production of α-globin usually due to the functional

deletion of 1 of the 4 α-globin genes. Based on the number of inherited alpha genes, alpha (+) thalassemia is subclassified into three general forms:

A. Thalassemia ($-\alpha/\alpha$ α) is characterized by inheritance of three normal α-globin genes. These patients are referred to clinically as silent carrier of alpha thalassemia. Other names for this condition are alpha thalassemia minima, alpha thalassemia-2 trait, and heterozygosity for alpha (+) thalassemia minor. The affected individuals exhibit no abnormality clinically and may be haematologically normal or have mild reductions in red cell mean corpuscular volume (MCV) and mean corpuscular haemoglobin (MCH).

B. Inheritance of two normal α-globin genes due to either heterozygosity for alpha (0) thalassemia (α $\alpha/--$) or homozygosity for alpha (+) thalassemia ($-\alpha/-\alpha$) results in the development of alpha thalassemia minor or alpha thalassemia-1 trait. The affected individuals are clinically normal but frequently have minimal anaemia and reduced mean corpuscular volume (MCV) and mean corpuscular haemoglobin (MCH). The red blood cell count is usually increased to over 5.5×10^{12} L^{-1}.

C. Inheritance of one normal gene ($-\alpha/--$) results in abundant formation of haemoglobin H composed of tetramers of excess β chains. This condition is known as HbH disease. The affected individuals have moderate to severe lifelong haemolytic anaemia, modest degrees of ineffective erythropoiesis, splenomegaly, and variable bony changes [5].

Case 4

Clinical presentation

- Mild features of anaemia

FBE RESULTS

CBC (Complete Blood Count) Parameters

Parameters	Results	Units
Hb	127	g/L
RCC	6.74	$\times 10^{12}\,L^{-1}$
PCV	0.44	
MCV	65	Fl
MCH	18.8	Pg
MCHC	290	g/L
RDW	19.1	%
Retics	N/A	%
WBC	6.0	$\times 10^{9}\,L^{-1}$
Plt count	226	$\times 10^{9}\,L^{-1}$

Differential White Cell Count

Cell Type	Percentages (%)	Absolute Count
Neutrophils	56	$3.36 \times 10^{9}\,L^{-1}$
Bands	0	$0 \times 10^{9}\,L^{-1}$
Lymphocytes	32	$1.92 \times 10^{9}\,L^{-1}$
Monocytes	8	$0.48 \times 10^{9}\,L^{-1}$
Eosinophils	2	$0.12 \times 10^{9}\,L^{-1}$
Basophils	2	$0.12 \times 10^{9}\,L^{-1}$
Metamyelocytes	0	
Myelocytes	0	

Continued

Differential White Cell Count—cont'd		
Cell Type	Percentages (%)	Absolute Count
Promyelocytes	0	
Blasts	0	
Prolymphocytes	0	
nRBC/100WBC	0	

RBC MORPHOLOGY

Moderate anisocytosis, moderate microcytosis, moderate hypochromia, occasional elongated cells, moderate target cells, and occasional basophilic stippling.

WBC MORPHOLOGY

Normal in number and morphology.

PLATELET MORPHOLOGY

Normal in number and morphology.

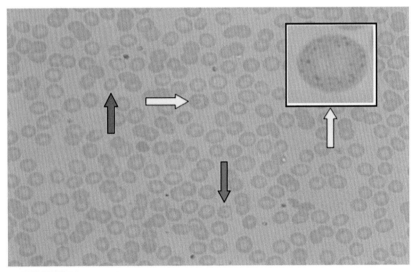

▬ Hypochromic, microcytic cell
▭ Target cell
▭ Basophilic stippling

Blood film image shows the most diagnostic features.

PROVISIONAL DIAGNOSIS

Based on clinical presentations, FBE results, and morphology, this case is highly suggestive of beta thalassemia minor and further tests are suggested.

DIFFERENTIAL DIAGNOSES

Other kind of thalassemia
Other haemoglobinopathies/unstable Hb levels
Anaemia of chronic disease
Sideroblastic anaemia
Iron deficiency anaemia

FURTHER TESTS AND EXPECTED RESULTS

Confirmation of Provisional Diagnosis

Haemoglobin electrophoresis and HPLC →

Hb A—decreased
Hb F—normal/slight increased
Hb A2—increased

Molecular tests → defect in β-globin gene

Exclusion of Differential Diagnoses

Iron profile

- Serum iron → normal
- TIBC → normal
- Ferritin → normal

FINAL DIAGNOSIS

Beta thalassemia minor is the final diagnosis for this case based on further tests that confirm the provisional diagnosis and exclude all the differential diagnoses.

PATHOPHYSIOLOGY OF BETA THALASSEMIA MINOR

Mutations in globin genes cause thalassemias. Beta thalassemia affects one or both of the β-globin genes. These mutations result in the impaired synthesis of the β-globin protein portion, a component of Hb, thus causing anaemia. The genetic defect usually is a missense or nonsense mutation in the β-globin gene, although occasional defects due to gene deletions of the β-globin gene and surrounding regions also have been reported.

In beta thalassemia minor (i.e. beta thalassemia trait or heterozygous carrier type), one of the β-globin genes is defective. The defect can be a complete absence of the β-globin protein (i.e. beta-zero thalassemia) or a reduced synthesis of the β-globin protein (i.e. beta-plus thalassemia) [6].

Case 5

Clinical presentations

- General features of anaemia.
- Pale, listless appearance, and jaundice.
- Slowed growth and delayed puberty.
- Enlarged spleen, liver, and heart. Bone problems (especially bones in the face).

FBE RESULTS

CBC (Complete Blood Count) Parameters

Parameters	Results	Units
Hb	41	g/L
RCC	1.50	$\times 10^{12}\,\text{L}^{-1}$
PCV	0.11	
MCV	73	Fl
MCH	27.0	Pg
MCHC	373	g/L
RDW	21.0	%
Retics	N/A	%
WBC	5.7	$\times 10^{9}\,\text{L}^{-1}$
Corrected WBC	0.9	$\times 10^{9}\,\text{L}^{-1}$
Plt count	328	$\times 10^{9}\,\text{L}^{-1}$

Differential White Cell Count

Cell Type	Percentages (%)	Absolute Count
Neutrophils	26	$0.23 \times 10^{9}\,\text{L}^{-1}$
Bands	3	$0.03 \times 10^{9}\,\text{L}^{-1}$
Lymphocytes	61	$0.55 \times 10^{9}\,\text{L}^{-1}$
Monocytes	7	$0.06 \times 10^{9}\,\text{L}^{-1}$
Eosinophils	2	$0.02 \times 10^{9}\,\text{L}^{-1}$
Basophils	1	$0.01 \times 10^{9}\,\text{L}^{-1}$

Continued

Differential White Cell Count—cont'd		
Cell Type	Percentages (%)	Absolute Count
Metamyelocytes	0	
Myelocytes	0	
Promyelocytes	0	
Blasts	0	
Prolymphocytes	0	
nRBC/100WBC	523	

RBC MORPHOLOGY

Marked anisocytosis, marked polychromasia, moderate target cells, moderate irregular contracted cells, moderate HJB, and marked nucleated RBC.

WBC MORPHOLOGY

Normal in number and morphology.

PLATELET MORPHOLOGY

Normal in number and morphology.

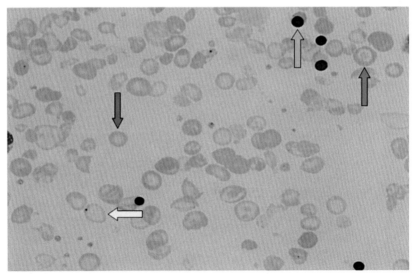

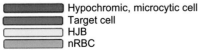 Hypochromic, microcytic cell
Target cell
HJB
nRBC

Blood film image shows the most diagnostic features.

PROVISIONAL DIAGNOSIS

Based on clinical presentations, FBE results, and morphology, this case is highly suggestive of beta thalassemia major and further tests are suggested.

DIFFERENTIAL DIAGNOSES

Other kind of thalassemia
Other haemoglobinopathies/unstable Hb levels
Anaemia of chronic disease
Sideroblastic anaemia
Some types of red cell membrane disorders

FURTHER TESTS AND EXPECTED RESULTS

Haemoglobin electrophoresis and HPLC →

Hb A—decreased/absent
Hb F—marked increased
Hb A2—variable

Iron profile

Serum iron → increased/normal
TIBC → decreased/normal
Ferritin → increased/normal
% saturation → increased

Molecular tests → defect in β-globin gene

FINAL DIAGNOSIS

Beta thalassemia major is the final diagnosis for this case based on further tests that confirm the provisional diagnosis and exclude all the differential diagnoses.

PATHOPHYSIOLOGY OF BETA THALASSEMIA MAJOR

Mutations in globin genes cause thalassemias. Beta thalassemia affects one or both of the β-globin genes. These mutations result in the impaired synthesis of the β-globin protein portion, a component of Hb, thus causing anaemia. The genetic defect usually is a missense or nonsense mutation in the β-globin gene, although occasional defects due to gene deletions of the β-globin gene and surrounding regions also have been reported.

In beta thalassemia major (i.e. homozygous beta thalassemia) the production of β-globin chains is severely impaired because both the β-globin genes are mutated. The severe imbalance of globin chain synthesis ($\alpha \gg \beta$) results in ineffective erythropoiesis and severe microcytic hypochromic anaemia.

The excess unpaired α-globin chains aggregate to form precipitates that damage red cell membranes, resulting in intravascular haemolysis. Premature destruction of erythroid precursors results in intramedullary death and ineffective erythropoiesis. The profound anaemia typically is associated with erythroid hyperplasia and extramedullary haematopoiesis [6].

Case 6

Clinical presentations

- General features of anaemia
- Flu-like symptoms and fever
- Hepatomegaly and splenomegaly

FBE RESULTS

CBC (Complete Blood Count) Parameters

Parameters	Results	Units
Hb	101	g/L
RCC	4.40	$\times 10^{12}$ L^{-1}
PCV	0.32	
MCV	73	Fl
MCH	23.0	Pg
MCHC	316	g/L
RDW	19.8	%
Retics	N/A	%
WBC	4.7	$\times 10^9$ L^{-1}
Plt count	195	$\times 10^9$ L^{-1}

Differential White Cell Count

Cell Type	Percentages (%)	Absolute Count
Neutrophils	31	1.46×10^9 L^{-1}
Bands	16	0.75×10^9 L^{-1}
Lymphocytes	36	1.69×10^9 L^{-1}
Monocytes	15	0.70×10^9 L^{-1}
Eosinophils	0	0×10^9 L^{-1}
Basophils	1	0.05×10^9 L^{-1}
Metamyelocytes	0	

Continued

Differential White Cell Count—cont'd		
Cell Type	Percentages (%)	Absolute Count
Myelocytes	1	
Promyelocytes	0	
Blasts	0	
Prolymphocytes	0	
nRBC/100WBC	0	

RBC MORPHOLOGY

Marked anisocytosis, moderate microcytes, mild hypochromia, mild target cells, moderate irregular contracted cells, occasional spherocytes, and mild elongated cells.

WBC MORPHOLOGY

Normal in number and morphology.

PLATELET MORPHOLOGY

Normal in number and morphology.

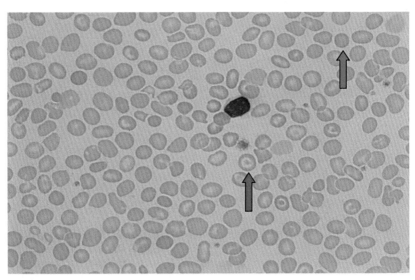

▭ Target cell
▭ Microsperocyte

Blood film image shows the most diagnostic features.

PROVISIONAL DIAGNOSIS

Based on clinical presentations, FBE results, and morphology, this case is highly suggestive of haemoglobin H disease and further tests are suggested.

DIFFERENTIAL DIAGNOSES

Beta thalassemia
Other haemolytic and dyserythropoietic anaemia
Congenital dyserythropoietic anaemia
Hereditary pyropoikilocytosis
Acquired Hb H disease
Iron deficiency anaemia
Autoimmune haemolytic anaemia
Nonimmune haemolytic anaemia

FURTHER TESTS AND EXPECTED RESULTS

Confirmation of Provisional Diagnosis

HbH Prep

Demonstration of Hb H inclusion

Hb-electrophoresis → presence of haemoglobin H.
HPLC → 2%–40% Hb H
Molecular tests → deletion of three α-globin genes.

Exclusion of Differential Diagnoses

Iron profile

Serum iron → normal
TIBC → normal
Ferritin → normal

DAT → negative

FINAL DIAGNOSIS

Haemoglobin H disease is the final diagnosis for this case based on further tests that confirm the provisional diagnosis and exclude all the differential diagnoses.
Pathophysiology (Refer to Pathophysiology of Case 3).

Case 7

Clinical presentations

- Mild features of anaemia
- Mild enlarged spleen
- Mild jaundice

FBE RESULTS

CBC (Complete Blood Count) Parameters

Parameters	Results	Units
Hb	99	g/L
RCC	4.37	$\times 10^{12}$ L^{-1}
PCV	0.30	
MCV	68	Fl
MCH	22.8	Pg
MCHC	324	g/L
RDW	19.0	%
Retics	N/A	%
WBC	9.0	$\times 10^9$ L^{-1}
Plt count	160	$\times 10^9$ L^{-1}

Differential White Cell Count

Cell Type	Percentages (%)	Absolute Count
Neutrophils	72	6.48×10^9 L^{-1}
Bands	2	0.18×10^9 L^{-1}
Lymphocytes	16	1.44×10^9 L^{-1}
Monocytes	7	0.63×10^9 L^{-1}
Eosinophils	2	0.18×10^9 L^{-1}
Basophils	1	0.1×10^9 L^{-1}
Metamyelocytes	0	

Differential White Cell Count—cont'd

Cell Type	Percentages (%)	Absolute Count
Myelocytes	0	
Promyelocytes	0	
Blasts	0	
Prolymphocytes	0	
nRBC/100WBC	0	

RBC MORPHOLOGY

Moderate anisocytosis, moderate microcytes, mild polychromasia, marked target cells, mild teardrop cells, mild irregular contracted cells, mild poikilocytosis, and occasional basophilic stippling.

WBC MORPHOLOGY

Normal in number and morphology.

PLATELET MORPHOLOGY

Normal in number and morphology.

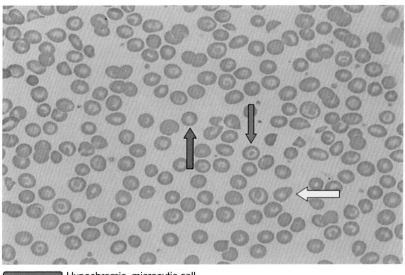

◼ Hypochromic, microcytic cell
◼ Target cell
◻ Teardrop cell

Blood film image shows the most diagnostic features.

PROVISIONAL DIAGNOSIS

Based on clinical presentations, FBE results, and morphology, this case is highly suggestive of haemoglobin E disease and further tests are suggested.

DIFFERENTIAL DIAGNOSES

Haemoglobin E/beta thalassemia
Beta thalassemia trait
Haemoglobin C disease
Iron deficiency anaemia

FURTHER TESTS AND EXPECTED RESULTS

Confirmation of Provisional Diagnosis

Haemoglobin electrophoresis and HPLC → mainly haemoglobin E and up to 5%–10% haemoglobin F (use acid pH to differentiate between Hb E & C)

Exclusion of Differential Diagnoses

Iron profile

Serum iron → normal
TIBC → normal
Ferritin → normal

FINAL DIAGNOSIS

Haemoglobin E disease is the final diagnosis for this case based on further tests that confirm the provisional diagnosis and exclude all the differential diagnoses.

PATHOPHYSIOLOGY OF HAEMOGLOBIN E DISEASE

Hb E is a β-chain variant in which lysine is substituted for glutamic acid in the 26th position. This substitution results in a net change in charge of +2, but the position of the substitution does not cause polymerization to occur [2].

Case 8

Clinical presentations

- Skin paleness
- Fatigue
- Dizziness
- Enlarged spleen and liver

FBE RESULTS

CBC (Complete Blood Count) Parameters

Parameters	Results	Units
Hb	99	g/L
RCC	4.42	$\times 10^{12}\,L^{-1}$
PCV	0.34	
MCV	79	Fl
MCH	22.0	Pg
MCHC	291	g/L
RDW	19.9	%
Retics	N/A	%
WBC	16.0	$\times 10^{9}\,L^{-1}$
Plt count	100	$\times 10^{9}\,L^{-1}$

Differential White Cell Count

Cell Type	Percentages (%)	Absolute Count
Neutrophils	90	$14.4 \times 10^{9}\,L^{-1}$
Bands	5	$0.8 \times 10^{9}\,L^{-1}$
Lymphocytes	5	$0.8 \times 10^{9}\,L^{-1}$
Monocytes	0	$0 \times 10^{9}\,L^{-1}$
Eosinophils	0	$0 \times 10^{9}\,L^{-1}$
Basophils	0	$0 \times 10^{9}\,L^{-1}$
Metamyelocytes	0	

Continued

Differential White Cell Count—cont'd		
Cell Type	Percentages (%)	Absolute Count
Myelocytes	0	
Promyelocytes	0	
Blasts	0	
Prolymphocytes	0	
nRBC/100WBC	0	

RBC MORPHOLOGY

Moderate anisocytosis, occasional macrocytes, occasional target cells, occasional Pappenheimer cells, and occasional nRBC.

WBC MORPHOLOGY

Moderate leucocytosis with marked neutrophilia.

PLATELET MORPHOLOGY

Mild thrombocytopenia with occasional giant platelets.

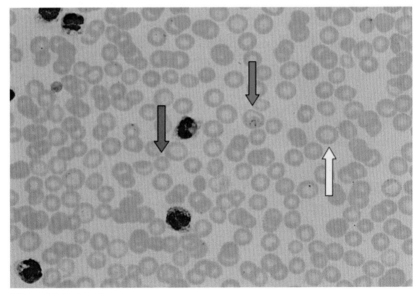

(Dimorphic features)

▭ Microcytes
▭ Pappenheimer body cell
▭ Macrocyte

Blood film image shows the most diagnostic features.

PROVISIONAL DIAGNOSIS

Based on clinical presentations, FBE results, and morphology, this case is highly suggestive of sideroblastic anaemia and further tests are suggested.

DIFFERENTIAL DIAGNOSES

Iron deficiency anaemia
Thalassemia
Anaemia of chronic disease
Haemoglobin C disease

FURTHER TESTS AND EXPECTED RESULTS

Confirmation of Provisional Diagnosis

Bone marrow test → sideroblast by using Prussian Blue stain

Iron profile

Serum iron → increased
TIBC → decreased/normal
Ferritin → increased
% transferrin saturation → increased/normal

Molecular tests → mutations of the ALAS-2 gene (in case of congenital sideroblastic anaemia)

Exclusion of Differential Diagnoses

HPLC and Hb electrophoresis → normal

FINAL DIAGNOSIS

Sideroblastic anaemia is the final diagnosis for this case based on further tests that confirm the provisional diagnosis and exclude all the differential diagnoses.

PATHOPHYSIOLOGY OF SIDEROBLASTIC ANAEMIA

Sideroblasts are not pathognomonic of any one disease but rather are a bone marrow manifestation of several diverse disorders. On a marrow stained with Prussian blue, a sideroblast is an erythroblast that has stainable deposits of iron in cytoplasm. When abundant, these deposits form a ring around the nucleus, and the cells become ring sideroblasts. Under normal circumstances, this iron would have been used to make haeme. The process only occurs in the bone marrow, because mature erythrocytes lack mitochondria, the nexus of haeme synthesis.

CONGENITAL SIDEROBLASTIC ANAEMIAS

Of the congenital sideroblastic anaemias, X-linked sideroblastic anaemias are further divided into pyridoxine-responsive and pyridoxine-resistant subtypes.

In the pyridoxine-responsive type of congenital sideroblastic anaemia, point mutations on the X chromosome have been identified that result in a δ-amino levulinic acid synthase (ALAS-2) with very low enzymatic activity. This impairs the first crucial step in the haeme synthesis pathway, the formation of δ-amino levulinic acid, resulting in anaemia despite intact iron delivery to the mitochondrion and with a lack of haeme in which iron is to be incorporated in the final step of this pathway.

A prototype of pyridoxine-resistant, X-linked sideroblastic anaemia is the ABC7 gene mutation. ABC-7 is an adenosine triphosphate (ATP)-dependent transporter protein involved in the cytosolic transfer of iron–sulphur complexes. In contrast to the pyridoxine-responsive sideroblastic anaemia, the ABC7 defect has a nonprogressive cerebellar ataxia component with diminished deep-tendon reflexes, incoordination, and elevated free erythrocyte protoporphyrin.

ACQUIRED SIDEROBLASTIC ANAEMIAS

Of the acquired sideroblastic anaemias, refractory anaemia with ring sidero-blasts (RARS) is a myelodysplastic syndrome characterized by an anaemia in which at least 15% of bone marrow erythroblasts are ringed sideroblasts, in addition to meeting other criteria for classification as a myelodysplastic syndrome. Approximately 15% of these patients also have thrombocytosis. Five per cent are JAK-2V617F – positive. RARS may also be considered a form of the myeloproliferative syndrome (MPS). Three to twelve per cent of patients with RARS progress to leukaemia. Interestingly with more numerous sideroblasts, there is lower the risk of progression. Ring sideroblasts, as a morphologic finding, may also be present in patients with other forms of the myelodysplastic syndrome.

Vitamin B6 (pyridoxine) forms pyridoxal phosphate, which acts as a coenzyme in the first, rate-limiting step in haeme formation catalysed by δ-ALAS. Deficiency of vitamin B6 causes sideroblastic anaemia.

Lead poisoning has been known to cause sideroblastic anaemia by inhibiting several enzymes involved in haeme synthesis, including δ-aminolevulinate dehydratase, coproporphyrin oxidase, and ferrochelatase.

Excessive alcohol consumption can cause several forms of anaemia by nutritional deficiencies (iron, folate), haemolysis, splenic sequestration due to liver cirrhosis, and sideroblastic anaemia by direct bone marrow toxicity to erythroid precursors, inhibition of pyridoxine, lead contamination of wine, and inhibition of ferrochelatase enzyme during haeme formation.

Drugs reported to cause sideroblastic anaemia include diverse classes such as antibiotics chloramphenicol, fusidic acid, linezolid, tetracycline, isoniazid;

hormones such as progesterone replacement; pain medicines like phenacetin; copper chelating agents such as penicillamine and trientine; and chemotherapy agents like busulfan. In most cases, stopping the drug reverses the sideroblastic changes.

Pure sideroblastic anaemia, although similar in some ways to RARS, differs from RARS by being less likely to transform to acute leukaemia [7].

Chapter 3

Normocytic Disorders

Case 9A

Clinical presentations

- Shortness of breath in altitude area
- Dizziness and headache
- Coldness in the hands and feet
- Pale skin and chest pain

FBE RESULTS

CBC (Complete Blood Count) Parameters		
Parameters	**Results**	**Units**
Hb	78	g/L
RCC	2.50	$\times 10^{12}\,L^{-1}$
PCV	0.23	
MCV	92	Fl
MCH	31.0	Pg
MCHC	339	g/L
RDW	19.8	%
Retics	N/A	%
WBC	6.5	$\times 10^{9}\,L^{-1}$
Plt count	400	$\times 10^{9}\,L^{-1}$

Haematology Case Studies with Blood Cell Morphology and Pathophysiology. http://dx.doi.org/10.1016/B978-0-12-811911-2.00003-9

Differential White Cell Count

Cell Type	Percentages (%)	Absolute Count
Neutrophils	52	$3.38 \times 10^9\ L^{-1}$
Bands	36	$2.34 \times 10^9\ L^{-1}$
Lymphocytes	36	$2.34 \times 10^9\ L^{-1}$
Monocytes	10	$0.65 \times 10^9\ L^{-1}$
Eosinophils	2	$0.13 \times 10^9\ L^{-1}$
Basophils	0	$0 \times 10^9\ L^{-1}$
Metamyelocytes	0	
Myelocytes	0	
Promyelocytes	0	
Blasts	0	
Prolymphocytes	0	
nRBC/100WBC	10	

RBC MORPHOLOGY

Marked anisocytosis, moderate polychromasia, moderate target cells, occasional elliptocytes, moderate sickle cells, moderate nRBC, and occasional HJB.

WBC MORPHOLOGY

Normal in number and morphology

PLATELET MORPHOLOGY

Normal in number with occasional platelets clumping

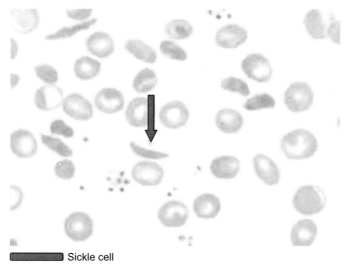

Sickle cell

Blood film image shows the most diagnostic features.

PROVISIONAL DIAGNOSIS

Based on clinical presentations, FBE results, and morphology, this case is highly suggestive of sickle cell anaemia and further tests are suggested.

DIFFERENTIAL DIAGNOSES

Combination of sickle cell and other haemoglobinopathies such as thalassemia and haemoglobin SC disease.
Liver and kidney disease.
Appendicitis.
Neurological disorders.
Other haemolytic anaemia.

FURTHER TESTS AND EXPECTED RESULTS

Sickle cell solubility test → positive
HPLC and Haemoglobin electrophoresis → 80% Hb S → acid to differentiate between Hb S & D.
Molecular tests → defect in the β-globin chain

FINAL DIAGNOSIS

Sickle cell anaemia is the final diagnosis for this case based on further tests that confirm the provisional diagnosis and exclude all the differential diagnoses.

PATHOPHYSIOLOGY OF SICKLE CELL ANAEMIA

Hb S arises from a mutation substituting thymine for adenine in the sixth codon of the beta-chain gene, GAG to GTG. This causes coding of valine instead of glutamic acid in position 6 of the Hb beta chain. The resulting Hb has the physical properties of forming polymers under deoxy conditions. It also exhibits changes in solubility and molecular stability. These properties are responsible for the profound clinical expressions of the sickling syndromes [8].

Case 9B

Clinical presentations

- Often kids with poor growth
- Fatigue
- Episodes of mild to severe pain
- Coldness in the hands and feet
- Pale skin and chest pain

FBE RESULTS

CBC (Complete Blood Count) Parameters

Parameters	Results	Units
Hb	76	g/L
RCC	3.43	$\times 10^{12}\,L^{-1}$
PCV	0.22	
MCV	65	Fl
MCH	22	Pg
MCHC	339	g/L
RDW	21	%
Retics	N/A	%
WBC	6.0	$\times 10^{9}\,L^{-1}$
Plt count	402	$\times 10^{9}\,L^{-1}$

Differential White Cell Count

Cell Type	Percentages (%)	Absolute Count
Neutrophils	34	$3.38 \times 10^{9}\,L^{-1}$
Bands		$2.34 \times 10^{9}\,L^{-1}$
Lymphocytes	55	$2.34 \times 10^{9}\,L^{-1}$
Monocytes	6	$0.65 \times 10^{9}\,L^{-1}$
Eosinophils	4	$0.13 \times 10^{9}\,L^{-1}$

Continued

Differential White Cell Count—cont'd		
Cell Type	Percentages (%)	Absolute Count
Basophils	1	$0 \times 10^9 \, L^{-1}$
Metamyelocytes	0	
Myelocytes	0	
Promyelocytes	0	
Blasts	0	
Prolymphocytes	0	
nRBC/100WBC	9	

RBC MORPHOLOGY

Marked anisocytosis, moderate microcytosis, hypochromia, target cells, occasional elliptocytes, moderate sickle cells, mild nRBC, and occasional HJB.

WBC MORPHOLOGY

Normal in morphology with lymphocytosis

PLATELET MORPHOLOGY

Normal in number and morphology

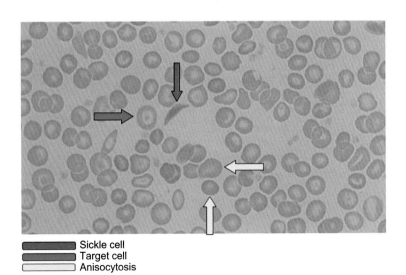

Sickle cell
Target cell
Anisocytosis

Blood film image shows the most diagnostic features.

PROVISIONAL DIAGNOSIS

Based on clinical presentations, FBE results, and morphology, this case is highly suggestive of sickle cell anaemia and possibly other haemoglobinopathy and further tests are suggested.

DIFFERENTIAL DIAGNOSES

Combination of sickle cell and other haemoglobinopathies such as thalassemia and haemoglobin SC disease.
Liver and kidney disease.
Appendicitis.
Neurological disorders.
Other haemolytic anaemia.

FURTHER TESTS AND EXPECTED RESULTS

Sickle cell solubility test → positive
HPLC and Haemoglobin electrophoresis → 6% HbA_2, 14% HbF & 80% Hb S → acid to differentiate between Hb S & D.
Molecular tests → multiple mutations HbS/β^0 thalassaemia

FINAL DIAGNOSIS

Sickle cell disease with thalassaemia is the final diagnosis for this case based on further tests that confirm the provisional diagnosis and exclude all the differential diagnoses.

PATHOPHYSIOLOGY OF SICKLE CELL DISEASE AND THALASSAEMIA

Sickle cell disease (SCD) and thalassaemia are recessively inherited genetic conditions, which affect the haemoglobin molecule. It is caused by errors in the genes for haemoglobin. The fault can either be structural such as SCD, or relate to an absence or reduction in globin chain synthesis, namely the thalassaemia syndromes [5, 6]. Both these pathologies are very complex and although they adversely affect the haemoglobin, they do so in very different ways. In patients with sickle cell disease (SCD) and β-thalassaemia, the gene that encodes for the production of adult haemoglobin is mutated. In SCD, red blood cells take on a characteristic sickle shape, stick together, and block the flow of blood and oxygen through small blood vessels resulting in painful complications. Patients with β-thalassaemia do not make red blood cells efficiently, requiring lifetime blood transfusions. While these patients' adult haemoglobin is defective, their foetal haemoglobin gene is perfectly normal and prevents both SCD and β-thalassaemia until it is switched off during development. Stimulating the switch from adult back to foetal haemoglobin to compensate for defective adult haemoglobin represents an attractive therapeutic strategy for both diseases.

Case 10

Clinical presentations

- Tiredness
- Pallor
- Breathlessness on exercise

FBE RESULTS

CBC (Complete Blood Count) Parameters

Parameters	Results	Units
Hb	112	g/L
RCC	3.59	$\times 10^{12}\,L^{-1}$
PCV	0.32	
MCV	91	Fl
MCH	31.3	Pg
MCHC	344	g/L
RDW	13.6	%
Retics	N/A	%
WBC	6.5	$\times 10^9\,L^{-1}$
Plt count	326	$\times 10^9\,L^{-1}$

Differential White Cell Count

Cell Type	Percentages (%)	Absolute Count
Neutrophils	72	$4.68 \times 10^9\,L^{-1}$
Bands	15	$0.97 \times 10^9\,L^{-1}$
Lymphocytes	6	$0.39 \times 10^9\,L^{-1}$
Monocytes	7	$0.45 \times 10^9\,L^{-1}$
Eosinophils	0	$0 \times 10^9\,L^{-1}$
Basophils	0	$0 \times 10^9\,L^{-1}$
Metamyelocytes	0	

Differential White Cell Count—cont'd

Cell Type	Percentages (%)	Absolute Count
Myelocytes	0	
Promyelocytes	0	
Blasts	0	
Prolymphocytes	0	
nRBC/100WBC	0	

RBC MORPHOLOGY

Normochromic cells, mild anisocytosis, mild poikilocytosis, and mild crenated cells.

WBC MORPHOLOGY

Normal in number with mild left shift.

PLATELET MORPHOLOGY

Normal in number and morphology.

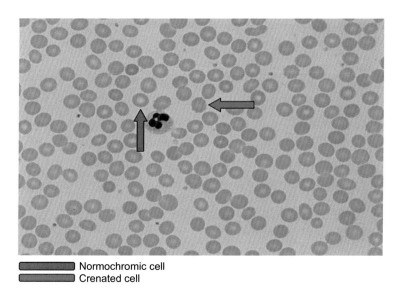

▭ Normochromic cell
▭ Crenated cell

Blood film image shows the most diagnostic features.

PROVISIONAL DIAGNOSIS

Based on clinical presentations, FBE results, and morphology, this case is highly suggestive of anaemia of chronic disease and further tests are suggested.

DIFFERENTIAL DIAGNOSES

Iron deficiency anaemia
Iron deficiency anaemia coexisting with ACD
Anaemia associated with chronic renal disease (erythropoietin deficiency)
Thalassemia
Anaemia due to drugs, radiation, and chemical exposure
Primary haematological disorder (e.g. myelodysplasia, multiple myeloma, leukaemia, lymphoma)

FURTHER TESTS AND EXPECTED RESULTS

Iron profile

Serum iron → decreased
TIBC → decreased
Ferritin → increased
% transferrin saturation → decreased
Transferrin receptor → normal
Bone marrow iron → present/increased

Biochemistry

Plasma viscosity → increased
C-reactive protein → increased
Serum albumin → reduced
Fibrinogen, alpha2 macroglobulin, and gamma globulins → increased

Exclude other differential diagnosis by doing special tests → the results should be normal

FINAL DIAGNOSIS

Anaemia of chronic disease is the final diagnosis for this case based on further tests that confirm the provisional diagnosis and exclude all the differential diagnoses.

PATHOPHYSIOLOGY OF ANAEMIA OF CHRONIC DISEASE

Anaemia of chronic illness traditionally encompassed any inflammatory, infectious, or malignant disease of a long-standing nature. The modern definition includes rheumatoid arthritis, severe trauma, heart disease, or diabetes mellitus. In these conditions, there is primarily a decreased availability of iron, relatively decreased levels of erythropoietin, and a mild decrease in the lifespan of RBCs to 70–80 days (normally 120 days).

Relatively recently, hepcidin, an endogenous antimicrobial peptide, has been identified. Secreted by the liver, hepcidin controls the level of plasma iron by regulating the intestinal absorption of dietary iron, as well as the release of iron from macrophages and the transfer of iron stored in the hepatocytes. Increase in hepcidin level in the course of inflammatory disease may be a significant mediator of the accompanying anaemia.

Another proposed mechanism for anaemia of chronic illness deals with cytokines, such as interleukins (IL-1 and IL-6), and tumour necrosis factor (TNF-alpha), which are believed to cause the destruction of RBC precursors and decrease the number of erythropoietin receptors on progenitor cells.

Whereas hypoxia in the individual with normal functioning kidneys leads to erythropoietin gene transcription, and hence increased RBC production, in those with anaemia of chronic kidney disease, there is primary deficiency of erythropoietin production by the interstitial fibroblasts, also known as type I interstitial cells, thereby leading to anaemia. The anaemia that develops is directly related to the amount of residual renal function. The kidneys are responsible for approximately 90% of erythropoietin production in an individual [9].

Case 11

Clinical presentations

- Fatigue and shortness of breath
- Chest pain
- Fever and flu-like illnesses
- Bleeding gums

FBE RESULTS

CBC (Complete Blood Count) Parameters

Parameters	Results	Units
Hb	100	g/L
RCC	3.40	$\times 10^{12}\,L^{-1}$
PCV	0.32	
MCV	94	Fl
MCH	29.0	Pg
MCHC	313	g/L
RDW	16.2	%
Retics	N/A	%
WBC	0.9	$\times 10^{9}\,L^{-1}$
Plt count	25	$\times 10^{9}\,L^{-1}$

Differential White Cell Count

Cell Type	Percentages (%)	Absolute Count
Neutrophils	30	$0.27 \times 10^{9}\,L^{-1}$
Bands	7	$0.06 \times 10^{9}\,L^{-1}$
Lymphocytes	37	$0.33 \times 10^{9}\,L^{-1}$
Monocytes	18	$0.16 \times 10^{9}\,L^{-1}$
Eosinophils	8	$0.07 \times 10^{9}\,L^{-1}$
Basophils	0	$0 \times 10^{9}\,L^{-1}$

Differential White Cell Count—cont'd

Cell Type	Percentages (%)	Absolute Count
Metamyelocytes	0	
Myelocytes	0	
Promyelocytes	0	
Blasts	0	
Prolymphocytes	0	
nRBC/100WBC	0	

RBC MORPHOLOGY

Normocytic normochromic cells and mild anisocytosis.

WBC MORPHOLOGY

Marked leukopenia with marked neutropenia.

PLATELET MORPHOLOGY

Marked thrombocytopenia with normal morphology.

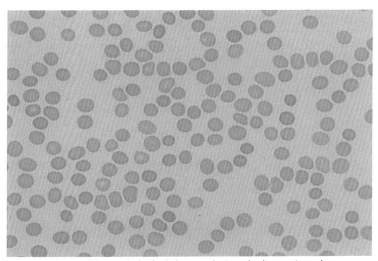

No specific morphological changes but marked pancytopenia

Blood film image shows the most diagnostic features.

PROVISIONAL DIAGNOSIS

Based on clinical presentations, FBE results, and morphology, this case is highly suggestive of aplastic anaemia and further tests are suggested.

DIFFERENTIAL DIAGNOSES

Acute lymphoblastic leukaemia
Acute myeloid leukaemia
Paroxysmal nocturnal haemoglobinuria
Myelodysplastic syndrome
Megaloblastic anaemia
Myelofibrosis
Lymphoma
Viral infection (HIV and EBV)
Thrombocytopenia
Fanconi anaemia

FURTHER TESTS AND EXPECTED RESULTS

Bone marrow test → marrow hypoplasia

Exclude other causes of marrow failure and pancytopenia

- Chest X-ray: infections
- Liver tests: liver diseases
- Viral studies: viral infections
- Vitamin B12 and folate levels: vitamin deficiency
- Blood tests for paroxysmal nocturnal haemoglobinuria
- Test for antibodies: immune competency
- X-rays, computed tomography (CT) scans, or ultrasound imaging tests: enlarged lymph nodes (sign of lymphoma), kidneys, and bones in arms and hands (abnormal in Fanconi anaemia)

FINAL DIAGNOSIS

Aplastic anaemia is the final diagnosis for this case based on further tests that confirm the provisional diagnosis and exclude all the differential diagnoses.

PATHOPHYSIOLOGY OF APLASTIC ANAEMIA

The primary lesion in acquired aplastic anaemia is a quantitative and qualitative deficiency of hematopoietic stem cells, rather than a defect of the bone marrow stroma or a deficiency in growth factors. The hematopoietic stem cells and early progenitor cell compartment is identified by expression of CD34 surface antigen. When measured by flow cytometry, the CD34 cell population in the bone marrow of patients with aplastic anaemia can be 10 times lower compared with normal individuals. There is an increase in apoptotic CD34 cells in aplastic anaemia, and they have an increased expression of Fas receptors that mediate apoptosis. Bone marrow cells in aplastic anaemia also have an increased expression of apoptotic genes determined by gene chip analysis. Individuals with aplastic anaemia have elevated levels of growth factors in their serum, such as erythropoietin. The pathophysiology of acquired aplastic anaemia involves the severe depletion of hematopoietic stem and progenitor cells from the bone marrow by a direct or indirect mechanism. In the direct mechanisms, a cytotoxic drug, chemical, radiation, or virus damages the DNA of the stem and progenitor cells causing apoptosis and cytolysis. In the indirect mechanisms, exposure to certain drugs or chemicals in susceptible individuals results in an autoimmune T-cell attack that destroys the stem and progenitor cells [2].

Case 12

Clinical presentations

- 32-year-old male had motor vehicle accident 3 months ago

FBE RESULTS

CBC (Complete Blood Count) Parameters

Parameters	Results	Units
Hb	135	g/L
RCC	4.5	$\times 10^{12} \, L^{-1}$
PCV	0.40	
MCV	89	Fl
MCH	30.0	Pg
MCHC	338	g/L
RDW	19.0	%
Retics	N/A	%
WBC	7.5	$\times 10^9 \, L^{-1}$
Plt count	180	$\times 10^9 \, L^{-1}$

Differential White Cell Count

Cell Type	Percentages (%)	Absolute Count
Neutrophils	48	$3.60 \times 10^9 \, L^{-1}$
Bands	0	$0 \times 10^9 \, L^{-1}$
Lymphocytes	46	$3.45 \times 10^9 \, L^{-1}$
Monocytes	4	$0.30 \times 10^9 \, L^{-1}$
Eosinophils	1	$0.07 \times 10^9 \, L^{-1}$
Basophils	1	$0.07 \times 10^9 \, L^{-1}$
Metamyelocytes	0	
Myelocytes	0	

Differential White Cell Count—cont'd		
Cell Type	Percentages (%)	Absolute Count
Promyelocytes	0	
Blasts	0	
Prolymphocytes	0	
nRBC/100WBC	0	

RBC MORPHOLOGY

Moderate anisocytosis, moderate poikilocytosis, mild polychromasia, mild target cells, occasional spherocyte, and marked Howell–Jolly body (HJB).

WBC MORPHOLOGY

Normal in number and morphology.

PLATELET MORPHOLOGY

Normal in number and morphology.

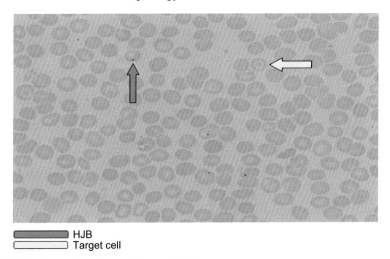

HJB
Target cell

Blood film image shows the most diagnostic features.

PROVISIONAL DIAGNOSIS

Based on clinical presentations, FBE results, and morphology, this case is highly suggestive of postsplenectomy change and further tests are suggested.

DIFFERENTIAL DIAGNOSES

Severe haemolytic anaemia
Megaloblastic anaemia
Hereditary spherocytosis

FURTHER TESTS AND EXPECTED RESULTS

Confirmation of Provisional Diagnosis

This case is highly diagnosed by clinical history and increased number of HJB in the peripheral blood smear and can be confirmed by exclusion of differential diagnosis that HJB cells can be seen in.

Exclusion of Differential Diagnoses

Vitamin B12 and folate → normal exclude megaloblastic anaemia
Acidified glycerol lysis test (AGLT) → negative exclude hereditary spherocytosis

FINAL DIAGNOSIS

Postsplenectomy change is the final diagnosis for this case based on further tests that confirm the provisional diagnosis and exclude all the differential diagnoses.

- **Alpha naphthyl acetate esterase (ANAE)** → negative

Immunophenotyping →

B (CD10, 22) & T (CD 2, 3, 5, 7) cell markers are negative
CD14&64 → negative

Cytogenetic and molecular study → t (15:17) and PML-RAR should be negative to exclude APML.

FINAL DIAGNOSIS

Acute myeloid leukaemia without maturation is the final diagnosis for this case based on further tests that confirm the provisional diagnosis and exclude all the differential diagnoses.

Chapter 4

Macrocytic Disorders

Case 13

Clinical presentations

- Dyspnoea
- Headache
- Fatigue
- Splenomegaly

FBE RESULTS

CBC (Complete Blood Count) Parameters

Parameters	Results	Units
Hb	114	g/L
RCC	3.29	$\times 10^{12} \, L^{-1}$
PCV	0.35	
MCV	107	Fl
MCH	35.0	Pg
MCHC	324	g/L
RDW	18.1	%
Retics	N/A	%
WBC	2.4	$\times 10^{9} \, L^{-1}$
Plt count	213	$\times 10^{9} \, L^{-1}$

Haematology Case Studies with Blood Cell Morphology and Pathophysiology. http://dx.doi.org/10.1016/B978-0-12-811911-2.00004-0

Differential White Cell Count

Cell Type	Percentages (%)	Absolute Count
Neutrophils	31	$0.74 \times 10^9 \, L^{-1}$
Bands	0	$0 \times 10^9 \, L^{-1}$
Lymphocytes	63	$1.51 \times 10^9 \, L^{-1}$
Monocytes	4	$0.10 \times 10^9 \, L^{-1}$
Eosinophils	1	$0.02 \times 10^9 \, L^{-1}$
Basophils	1	$0.02 \times 10^9 \, L^{-1}$
Metamyelocytes	0	
Myelocytes	0	
Promyelocytes	0	
Blasts	0	
Prolymphocytes	0	
nRBC/100WBC	0	

RBC MORPHOLOGY

Moderate anisocytosis, moderate macrospherocytes.

WBC MORPHOLOGY

Mild leucopenia and mild neutropenia.

PLATELET MORPHOLOGY

Normal in number and morphology.

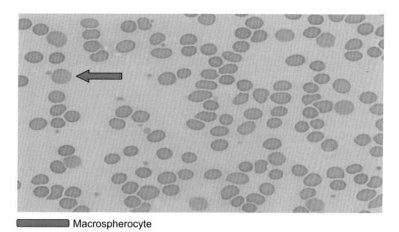

▬▬▬ Macrospherocyte

Blood film image shows the most diagnostic features.

PROVISIONAL DIAGNOSIS

Based on clinical presentations, FBE results, and morphology, this case is highly suggestive of macrocytosis and further tests are suggested.

DIFFERENTIAL DIAGNOSES

Megaloblastic anaemia
Liver disease
Drug-induced haemolytic anaemia (DIHA)
Myelodysplastic syndrome (MDS)
Hypothyroidism
Autoimmune haemolytic anaemia (AIHA)

FURTHER TESTS AND EXPECTED RESULTS

Macrocytosis can be caused by many conditions and all of them are included in the differential diagnosis.

Serum vitamin B12, serum folate, and red cell folate → decreased in megaloblastic anaemia
Serum antibody to intrinsic factor → positive (in case of pernicious anaemia)
Liver function tests → abnormal in case of liver disease
Drug history → positive in DIHA
DAT → positive in AIHA
Cytogenetic abnormality such as del 5 q → positive in case of MDS

FINAL DIAGNOSIS

Macrocytosis is the final diagnosis for this case based on further tests that confirm the provisional diagnosis and exclude all the differential diagnoses.

PATHOPHYSIOLOGY OF MACROCYTOSIS

The most common cause of macrocytic anaemia is megaloblastic anaemia, which is the result of impaired DNA synthesis. Although DNA synthesis is impaired, RNA synthesis is unaffected, leading to a build-up of cytoplasmic components in a slowly dividing cell. This results in a larger-than-normal cell. The nuclear chromatin of these cells also has an altered appearance.

Hydroxyurea, an agent now commonly used to decrease the number of vaso-occlusive pain crises in patients with sickle cell disease, interferes with DNA synthesis, causing a macrocytosis by which compliance with therapy may be monitored. Patient compliance with zidovudine, an agent used in the treatment of patients with HIV, may be monitored in the same way.

Nonmegaloblastic macrocytic anaemias are those in which no impairment of DNA synthesis occurs. Included in this category are disorders associated with increased membrane surface area, accelerated erythropoiesis, alcoholism, and chronic obstructive pulmonary disease (COPD). The macrocytosis associated with COPD is attributed to excess cell water that is secondary to carbon dioxide retention.

Patients with hepatic disease and obstructive jaundice have a macrocytosis that is secondary to increased cholesterol and/or phospholipids deposited on the membranes of circulating RBCs. Similarly, in splenectomized patients, RBC membrane lipids that usually are removed during maturation in the spleen are not effectively removed, leading to a larger-than-normal cell.

In patients with haemolytic anaemia or posthaemorrhagic anaemia, the reticulocyte count increases. The reticulocyte, an immature RBC, is approximately 20% larger than the more mature RBC. When released prematurely from the marrow, the volume of the reticulocyte is averaged with the volume of the more mature RBC, and the resultant MCV is increased.

Macrocytosis, sometimes without associated anaemia, is often evident in persons with chronic alcoholism. Although the macrocytosis of alcoholism may be secondary to poor nutrition with a resulting folate or vitamin B12 deficiency, it is more often due to a direct toxicity of the alcohol on the marrow. The macrocytosis of alcoholism usually reverses only after months of abstinence from alcohol [10].

Case 14

Clinical presentations

- Numbness or tingling in hands and feet
- Difficulty walking and weak muscles
- Nausea
- Lack of energy or tiring easily (fatigue)
- Increased heart rate (tachycardia)

FBE RESULTS

CBC (Complete Blood Count) Parameters

Parameters	Results	Units
Hb	69	g/L
RCC	2.01	$\times 10^{12}$ L^{-1}
PCV	0.22	
MCV	109	Fl
MCH	34.3	Pg
MCHC	313	g/L
RDW	23.0	%
Retics	N/A	%
WBC	9.0	$\times 10^9$ L^{-1}
Plt count	250	$\times 10^9$ L^{-1}

Differential White Cell Count

Cell Type	Percentages (%)	Absolute Count
Neutrophils	75	6.75×10^9 L^{-1}
Bands	0	0×10^9 L^{-1}
Lymphocytes	20	1.8×10^9 L^{-1}
Monocytes	1	0.09×10^9 L^{-1}
Eosinophils	4	0.36×10^9 L^{-1}

Continued

Differential White Cell Count—cont'd		
Cell Type	Percentages (%)	Absolute Count
Basophils	0	$0 \times 10^9 \, L^{-1}$
Metamyelocytes	0	
Myelocytes	0	
Promyelocytes	0	
Blasts	0	
Prolymphocytes	0	
nRBC/100WBC	2	

RBC MORPHOLOGY

Marked anisocytosis, moderate macrocytes, occasional teardrop cells, moderate fragment cells, and moderate ovalocytes.

WBC MORPHOLOGY

Normal in number with marked hypersegmented neutrophils.

PLATELET MORPHOLOGY

Normal in number and morphology.

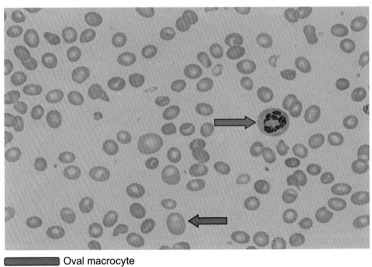

▬ Oval macrocyte
▬ Hypersegmented neutrophil

Blood film image shows the most diagnostic features.

PROVISIONAL DIAGNOSIS

Based on clinical presentations, FBE results, and morphology, this case is highly suggestive of megaloblastic anaemia and further tests are suggested.

DIFFERENTIAL DIAGNOSES

Liver disease
Drug-induced haemolytic anaemia (DIHA)
Myelodysplastic syndrome (MDS)

FURTHER TESTS AND EXPECTED RESULTS

Confirmation of Provisional Diagnosis

Serum vitamin B12 → decreased
Serum folate → decreased
Red cell folate → decreased
Serum antibody to intrinsic factor → positive (in case of pernicious anaemia)
Bone marrow test → hypercellular

Exclusion of Differential Diagnoses

Liver function tests → normal
Drug history → no drugs used

FINAL DIAGNOSIS

Megaloblastic anaemia is the final diagnosis for this case based on further tests that confirm the provisional diagnosis and exclude all the differential diagnoses.

PATHOPHYSIOLOGY OF MEGALOBLASTIC ANAEMIA

The molecular basis for megaloblastosis is a failure in the synthesis and assembly of DNA. The most common causes of megaloblastosis are cobalamin and folate deficiencies. Cobalamin metabolism and folate metabolism are intricately related, and abnormalities in these pathways are believed to lead to the attenuated production of DNA.

Methotrexate-induced megaloblastosis has been ascribed to a deficiency in deoxythymidine triphosphate (dTTP) that is consumed by the methyl folate trap. Evidence exists that megaloblastosis is caused by interference of folate metabolism by the inhibition of methionine synthesis. However, because of dietary folate deficiency, the size of the dTTP pool is normal or increased in persons with megaloblastosis.

Impairment in the deoxyuridine monophosphate (dUMP) and deoxythymidine monophosphate (dTMP) pathway may be responsible for nutritional megaloblastosis.

A hallmark of megaloblastic anaemia is ineffective erythropoiesis, as evidenced by erythroid hyperplasia in the bone marrow, a decreased peripheral reticulocyte count, and an elevation in lactate dehydrogenase (LDH) and indirect bilirubin levels. The pathogenesis of these findings is the intramedullary destruction of fragile and abnormal megaloblastic erythroid precursors.

An understanding of the source of cobalamin and folate is important to understand the pathogenesis of the development of megaloblastosis. Dietary intake is the source of cobalamin and folate because humans cannot synthesize these substances. Cobalamin must be bound to intrinsic factor (IF), and this complex is taken up in the terminal ileum. Once absorbed, cobalamin is bound to another protein, transcobalamin II (TCII), and is transported to storage sites. Abnormalities in any of these steps in cobalamin transport can lead to deficiencies in this substance. Considerable amounts of cobalamin are accumulated in storage sites; this explains why years elapse before cobalamin deficiency develops in patients who cannot take up dietary cobalamin.

Although the processing and transport of ingested folate is complex, folate-induced megaloblastosis is rarely caused by abnormalities in transport but instead is most often caused by dietary insufficiency. Folate deficiency can be caused by malabsorption in patients with sprue. In contrast to cobalamin, very little folate is stored; this explains why folate deficiency can occur within months of cessation of folate ingestion.

Megaloblastosis can also be caused by disorders in which cobalamin and folate uptake and metabolism are not affected [11].

Case 15

Clinical presentations

- Pale and yellow skin
- Redness and itchiness of eyes
- Headache
- Loss of appetite
- Hepatomegaly

FBE RESULTS

CBC (Complete Blood Count) Parameters

Parameters	Results	Units
Hb	110	g/L
RCC	2.97	$\times 10^{12}\,L^{-1}$
PCV	0.33	
MCV	110	Fl
MCH	37.0	Pg
MCHC	336	g/L
RDW	N/A	%
Retics	N/A	%
WBC	15.0	$\times 10^{9}\,L^{-1}$
Plt count	111	$\times 10^{9}\,L^{-1}$

Differential White Cell Count

Cell Type	Percentages (%)	Absolute Count
Neutrophils	79	$11.85 \times 10^{9}\,L^{-1}$
Bands	10	$1.5 \times 10^{9}\,L^{-1}$
Lymphocytes	5	$0.75 \times 10^{9}\,L^{-1}$
Monocytes	5	$0.75 \times 10^{9}\,L^{-1}$
Eosinophils	1	$0.15 \times 10^{9}\,L^{-1}$

Continued

Differential White Cell Count—cont'd		
Cell Type	Percentages (%)	Absolute Count
Basophils	0	$0 \times 10^9 \, L^{-1}$
Metamyelocytes	0	
Myelocytes	0	
Promyelocytes	0	
Blasts	0	
Prolymphocytes	0	
nRBC/100WBC	0	

RBC MORPHOLOGY

Moderate macrocytosis, marked target cells, and occasional rouleaux.

WBC MORPHOLOGY

Mild leucocytosis with moderate neutrophilia and mild left shift.

PLATELET MORPHOLOGY

Mild thrombocytopenia with normal morphology.

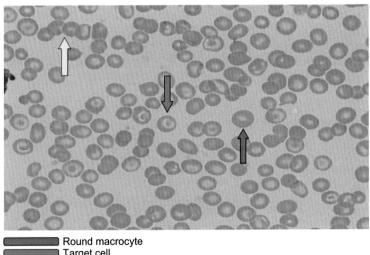

Round macrocyte
Target cell
Rouleaux

Blood film image shows the most diagnostic features.

PROVISIONAL DIAGNOSIS

Based on clinical presentations, FBE results, and morphology, this case is highly suggestive of liver disease and further tests are suggested.

DIFFERENTIAL DIAGNOSES

Megaloblastic anaemia
Drug-induced haemolytic anaemia (DIHA)
Liver cancer

FURTHER TESTS AND EXPECTED RESULTS

Liver function test → abnormal
Exclude other causes of macrocytosis → they should be normal

FINAL DIAGNOSIS

Liver disease is the final diagnosis for this case based on further tests that confirm the provisional diagnosis and exclude all the differential diagnoses.

PATHOPHYSIOLOGY OF MACROCYTOSIS CAUSED BY LIVER DISEASE

Patients with hepatic disease and obstructive jaundice have a macrocytosis that is secondary to increased cholesterol and/or phospholipids deposited on the membranes of circulating RBCs [10].

Chapter 5

Nonimmune Haemolytic Disorders (RBC Metabolic Abnormalities)

Case 16

Clinical presentations

- Enlarged spleen
- Yellow skin and eyes
- General weakness, fatigue, and shortness of breath
- Abdominal pain
- Itching

FBE RESULTS

CBC (Complete Blood Count) Parameters

Parameters	Results	Units
Hb	92	g/L
RCC	2.90	$\times 10^{12} \, L^{-1}$
PCV	0.26	
MCV	89	Fl
MCH	31.7	Pg
MCHC	360	g/L
RDW	17.3	%
Retics	N/A	%
WBC	13.5	$\times 10^{9} \, L^{-1}$
Plt count	250	$\times 10^{9} \, L^{-1}$

Haematology Case Studies with Blood Cell Morphology and Pathophysiology. http://dx.doi.org/10.1016/B978-0-12-811911-2.00005-2

Differential White Cell Count

Cell Type	Percentages (%)	Absolute Count
Neutrophils	64	$8.64 \times 10^9 \, L^{-1}$
Bands	0	$0 \times 10^9 \, L^{-1}$
Lymphocytes	35	$4.72 \times 10^9 \, L^{-1}$
Monocytes	1	$0.13 \times 10^9 \, L^{-1}$
Eosinophils	0	$0 \times 10^9 \, L^{-1}$
Basophils	0	$0 \times 10^9 \, L^{-1}$
Metamyelocytes	0	
Myelocytes	0	
Promyelocytes	0	
Blasts	0	
Prolymphocytes	0	
nRBC/100WBC	0	

RBC MORPHOLOGY

Moderate anisocytosis, moderate polychromasia, moderate blister cells, occasional crenated cells, and moderate bite cells.

WBC MORPHOLOGY

Mild leucocytosis with mild neutrophilia and mild lymphocytosis.

PLATELET MORPHOLOGY

Normal in number and morphology.

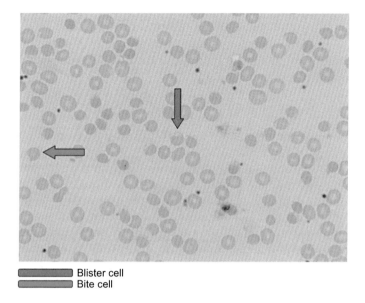

Blister cell
Bite cell

Blood film image shows the most diagnostic features.

PROVISIONAL DIAGNOSIS

Based on clinical presentations, FBE results, and morphology, this case is highly suggestive of glucose-6-phosphate dehydrogenase deficiency and further tests are suggested.

DIFFERENTIAL DIAGNOSES

Sickle cell anaemia
Hereditary spherocytosis
Drug-induced haemolytic anaemia

FURTHER TESTS AND EXPECTED RESULTS

- **Glucose fluorescent spot test** → no fluorescence
- **G6PD dye reduction test** → no colour change after 1 hour
- **Molecular tests** → mutation in G6PD gene
- **G6PD enzyme assay** → decreased
- **Direct Coombs' test** → negative
- **Exclude other causes of haemolysis**

FINAL DIAGNOSIS

Glucose-6-phosphate dehydrogenase deficiency is the final diagnosis for this case based on further tests that confirm the provisional diagnosis and exclude all the differential diagnoses.

PATHOPHYSIOLOGY OF G6PD DEFICIENCY

The enzyme G6PD reduces nicotinamide adenine dinucleotide phosphate (NADP) while oxidizing glucose-6-phosphate. It governs the rate of reduction of NADP which restores reduced glutathione (GSH) through H+ transfer by glutathione reductase in the hexose monophosphate shunt. G6PD provides the only means of generating the reduced form of nicotinamide adenine dinucleotide phosphate (NADPH), and in its absence the erythrocyte is particularly vulnerable to oxidative damage. Most oxidant drugs enable interaction between molecular oxygen and components of the RBC. RBCs with normal G6PD activity are able to detoxify the oxidative compounds and safeguard the haemoglobin. In patients with deficient G6PD activity, the exposure of RBCs to oxidative agents causes oxidation of membrane thiols, which produces several consequences, the most conspicuous being the appearance of Heinz bodies (intracellular irreversible precipitates of denatures haemoglobin) along with skeletal abnormalities, causing K+ and N+ leak rate to increase. Coagulum haemoglobin separates from the RBC membrane leaving a clear area similar to a bite being taken out of a cell and these cells can be seen in the peripheral blood [2].

Case 17

Clinical presentations

- General features of anaemia
- Yellow skin and eyes
- Fatigue and lethargy
- Pallor

FBE RESULTS

CBC (Complete Blood Count) Parameters

Parameters	Results	Units
Hb	92	g/L
RCC	2.90	$\times 10^{12}\,L^{-1}$
PCV	0.26	
MCV	89	Fl
MCH	31.7	Pg
MCHC	360	g/L
RDW	17.3	%
Retics	N/A	%
WBC	13.5	$\times 10^{9}\,L^{-1}$
Plt count	250	$\times 10^{9}\,L^{-1}$

Differential White Cell Count

Cell Type	Percentages (%)	Absolute Count
Neutrophils	59	$10.09 \times 10^{9}\,L^{-1}$
Bands	2	$0.34 \times 10^{9}\,L^{-1}$
Lymphocytes	19	$3.25 \times 10^{9}\,L^{-1}$
Monocytes	14	$2.39 \times 10^{9}\,L^{-1}$
Eosinophils	4	$0.68 \times 10^{9}\,L^{-1}$
Basophils	1	$0.17 \times 10^{9}\,L^{-1}$

Continued

Differential White Cell Count—cont'd		
Cell Type	Percentages (%)	Absolute Count
Metamyelocytes	1	
Myelocytes	0	
Promyelocytes	0	
Blasts	0	
Prolymphocytes	0	
nRBC/100WBC	0	

RBC MORPHOLOGY

Marked anisocytosis, moderate macrocytes, marked polychromasia, mild target cells, mild spherocytes, moderate irregular contracted cells, marked spur cells, marked Pappenheimer cells, and mild HJB.

WBC MORPHOLOGY

Mild leucocytosis with mild neutrophilia.

PLATELET MORPHOLOGY

Normal in number and morphology.

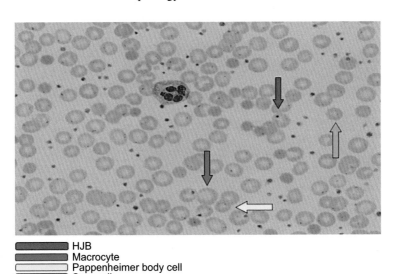

◼ HJB
◼ Macrocyte
◻ Pappenheimer body cell
◼ Spur cell

Blood film image shows the most diagnostic features.

PROVISIONAL DIAGNOSIS

Based on clinical presentations, FBE results, and morphology, this case is highly suggestive of pyruvate kinase deficiency and further tests are suggested.

DIFFERENTIAL DIAGNOSES

Autoimmune haemolytic anaemia (AIHA)
Thalassemia
Haemoglobin E disease

FURTHER TESTS AND EXPECTED RESULTS

Confirmation of Provisional Diagnosis

Haemoglobin electrophoresis → normal
Osmotic fragility → normal
Enzyme assay → deficient
Molecular tests → mutation in PK gene

Exclusion of Differential Diagnoses

HPLC and Hb electrophoresis → normal
DAT → negative

FINAL DIAGNOSIS

Pyruvate kinase deficiency is the final diagnosis for this case based on further tests that confirm the provisional diagnosis and exclude all the differential diagnoses.

PATHOPHYSIOLOGY OF PK DEFICIENCY

PK is a rate-limiting key enzyme of the glycolytic pathway. It catalyses the conversion of phosphoenolpyruvate to pyruvate with regeneration of ATP. The ATP content is often decreased, as are the adenosine diphosphate and adenosine monophosphate contents of the RBC. The 2,3-BPG is often increased approximately twofold. These alterations in the PK-deficient RBCs result in a rigid cell that is removed by the macrophages of spleen and liver [2].

Chapter 6

Nonimmune Haemolytic Disorders (RBC Membrane Abnormalities)

Case 18

Clinical presentations

- Fatigue
- Pallor
- Jaundice
- Splenomegaly

FBE RESULTS

CBC (Complete Blood Count) Parameters

Parameters	Results	Units
Hb	98	g/L
RCC	3.62	$\times 10^{12}\,L^{-1}$
PCV	0.30	
MCV	83	Fl
MCH	27.1	Pg
MCHC	327	g/L
RDW	19.9	%
Retics	N/A	%
WBC	5.5	$\times 10^{9}\,L^{-1}$
Plt count	300	$\times 10^{9}\,L^{-1}$

Differential White Cell Count

Cell Type	Percentages (%)	Absolute Count
Neutrophils	62	$3.41 \times 10^9 \, L^{-1}$
Bands	2	$0.11 \times 10^9 \, L^{-1}$
Lymphocytes	28	$1.54 \times 10^9 \, L^{-1}$
Monocytes	5	$0.27 \times 10^9 \, L^{-1}$
Eosinophils	3	$0.16 \times 10^9 \, L^{-1}$
Basophils	0	$0 \times 10^9 \, L^{-1}$
Metamyelocytes	0	
Myelocytes	0	
Promyelocytes	0	
Blasts	0	
Prolymphocytes	0	
nRBC/100WBC	0	

RBC MORPHOLOGY

Moderate anisocytosis, marked polychromasia, marked spherocytes, and moderate crenated cells.

WBC MORPHOLOGY

Normal in number and morphology.

PLATELET MORPHOLOGY

Normal in number and morphology.

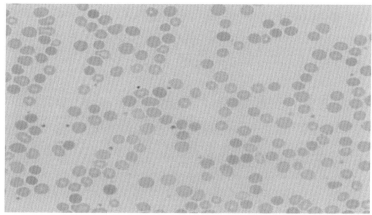

The most diagnostic features; marked spherocytosis

Blood film image shows the most diagnostic features.

PROVISIONAL DIAGNOSIS

Based on clinical presentations, FBE results, and morphology, this case is highly suggestive of hereditary spherocytosis in crisis and further tests are suggested.

DIFFERENTIAL DIAGNOSES

Autoimmune haemolytic anaemia (AIHA)
Drug induced haemolytic anaemia (DIHA)
Severe burns
Septicaemia
Biliary disease

FURTHER TESTS AND EXPECTED RESULTS

Biochemistry → increase in unconjugated bilirubin
Immunohaematology—DAT → negative
Osmotic fragility test → positive (shift to the right)
Acidified glycerol lysis test (AGLT) → positive
Pink test → positive
Flowcytometric analysis → Decreased fluorescence of band 3 protein
Molecular tests → defect in alpha and beta spectrin

FINAL DIAGNOSIS

Hereditary spherocytosis in crisis is the final diagnosis for this case based on further tests that confirm the provisional diagnosis and exclude all the differential diagnoses.

PATHOPHYSIOLOGY OF HEREDITARY SPHEROCYTOSIS

The primary molecular defects in hereditary spherocytosis (HS) are in the membrane skeletal proteins, specially the proteins that connect the membrane skeleton to the lipid bilayer. These include spectrin, ankyrin, protein 4.2, and band 3. The degree of deficiency correlates with the severity of the disease and with the degree of spherocytosis measured by osmotic fragility test. Heterozygous alpha-spectrin defects are usually asymptomatic but the combination of two alpha-spectrin gene defects in Trans leads to a significant alpha-spectrin deficiency and severe spherocytic anaemia. Several mutations have been identified for beta-spectrin defects that cause dominant HS. Biochemical analyses and genetic studies have implicated ankyrin defects in many cases of HS. In about one-third of patients with HS, the primary protein defect is in band 3 and these patients may have a combined deficiency of band 3 and protein 4.2. The membrane skeletal protein causes RBCs progressively to lose unsupported lipid membrane because of local disconnection of the skeletal and bilayer. Essentially, small proteins of the membrane peel off without loss of much volume. They acquire a decreased surface area-to-volume ratio and a spheroidal shape on the blood smear. These cells are rigid and are not as deformable as normal biconcave disc RBCs, and their survival in the spleen is decreased. The spleen selectively sequesters spherocytes from HS as they try to squeeze through spaces in the endothelial cells of the venous sinuses. As these spherocytes move more slowly through the narrow, elliptical fenestration of the splenic sinusoids, which are smaller than RBCs, they are especially susceptible to even further membrane loss and become so damaged that they are selectively removed by the macrophages of the spleen [2].

Case 19

Clinical presentations

● Fatigue
● Pallor
● Jaundice
● Splenomegaly

FBE RESULTS

CBC (Complete Blood Count) Parameters

Parameters	Results	Units
Hb	100	g/L
RCC	3.33	$\times 10^{12}\,L^{-1}$
PCV	0.30	
MCV	90	Fl
MCH	30.0	Pg
MCHC	333	g/L
RDW	14.0	%
Retics	N/A	%
WBC	9.2	$\times 10^{9}\,L^{-1}$
Plt count	200	$\times 10^{9}\,L^{-1}$

Differential White Cell Count

Cell Type	Percentages (%)	Absolute Count
Neutrophils	70	$6.44 \times 10^{9}\,L^{-1}$
Bands	6	$0.55 \times 10^{9}\,L^{-1}$
Lymphocytes	23	$2.15 \times 10^{9}\,L^{-1}$
Monocytes	1	$0.09 \times 10^{9}\,L^{-1}$
Eosinophils	0	$0 \times 10^{9}\,L^{-1}$
Basophils	0	$0 \times 10^{9}\,L^{-1}$

Continued

Differential White Cell Count—cont'd		
Cell Type	Percentages (%)	Absolute Count
Metamyelocytes	0	
Myelocytes	0	
Promyelocytes	0	
Blasts	0	
Prolymphocytes	0	
nRBC/100WBC	0	

RBC MORPHOLOGY

Marked elliptocytosis.

WBC MORPHOLOGY

Normal in number and morphology.

PLATELET MORPHOLOGY

Normal in number and morphology.

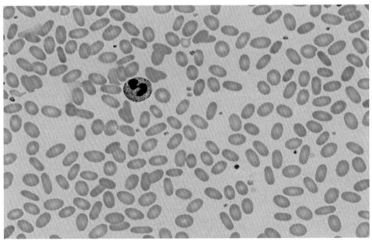

The most diagnostic features; marked elliptocytosis

Blood film image shows the most diagnostic features.

PROVISIONAL DIAGNOSIS

Based on clinical presentations, FBE results, and morphology, this case is highly suggestive of hereditary elliptocytosis and further tests are suggested.

DIFFERENTIAL DIAGNOSES

Iron deficiency anaemia
Megaloblastic anaemia
Hereditary pyropoikilocytosis
Myelofibrosis
Myelodysplasia
Pyruvate kinase deficiency

FURTHER TESTS AND EXPECTED RESULTS

Biochemistry → increase in unconjugated bilirubin and LDH
Osmotic fragility test → positive
DAT → negative
Molecular genetic testing for the presence of mutations in the specific protein molecules of the RBCs → positive (mutation of alpha and beta spectrin)

FINAL DIAGNOSIS

Hereditary elliptocytosis is the final diagnosis for this case based on further tests that confirm the provisional diagnosis and exclude all the differential diagnoses.

PATHOPHYSIOLOGY OF HEREDITARY ELLIPTOCYTOSIS

Abnormalities of either alpha-spectrin or beta-spectrin are associated with hereditary elliptocytosis (HE). The type of structural change in the membrane protein spectrin seems to be associated with the clinical manifestations. Protein 4.1 deficiencies also have been linked to HE. The defects affect the RBCs deformability and reflect the failure of RBCs to return to their normal disc shape after being deformed by the shear forces in the microcirculation. Many patients with common HE have an abnormality of the spectrin heterodimer head region, which cause an inability to associate the dimer form to tetramers and high-order oligomers and these abnormalities commonly involve the N-terminal peptide of alpha chain spectrin. Defects in 4.1 proteins also are a common cause of elliptocytosis in some Arabic and European populations. HE patients all show decreased RBC thermal stability in varying degrees. This fragile self-association of spectrin to spectrin weakens the membrane skeleton of these HE cells and diminishes their resistance to shear stress [2].

Case 20

Clinical presentations

- Pallor
- Fatigue
- Splenomegaly
- Abdominal pain

FBE RESULTS

CBC (Complete Blood Count) Parameters

Parameters	Results	Units
Hb	120	g/L
RCC	4.30	$\times 10^{12}\,L^{-1}$
PCV	0.37	
MCV	86	Fl
MCH	28.0	Pg
MCHC	324	g/L
RDW	17.2	%
Retics	N/A	%
WBC	6.1	$\times 10^9\,L^{-1}$
Plt count	150	$\times 10^9\,L^{-1}$

Differential White Cell Count

Cell Type	Percentages (%)	Absolute Count
Neutrophils	55	$3.35 \times 10^9\,L^{-1}$
Bands	0	$0 \times 10^9\,L^{-1}$
Lymphocytes	38	$2.32 \times 10^9\,L^{-1}$
Monocytes	4	$0.24 \times 10^9\,L^{-1}$
Eosinophils	1	$0.06 \times 10^9\,L^{-1}$
Basophils	2	$0.12 \times 10^9\,L^{-1}$

Differential White Cell Count—cont'd

Cell Type	Percentages (%)	Absolute Count
Metamyelocytes	0	
Myelocytes	0	
Promyelocytes	0	
Blasts	0	
Prolymphocytes	0	
nRBC/100WBC	0	

RBC MORPHOLOGY

Moderate anisocytosis, mild polychromasia, moderate irregular contracted cells, mild blister cells, moderate target cells, and mild stomatocytes.

WBC MORPHOLOGY

Normal in number and morphology.

PLATELET MORPHOLOGY

Normal in number and morphology.

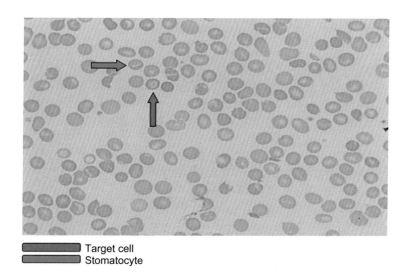

■ Target cell
■ Stomatocyte

Blood film image shows the most diagnostic features.

PROVISIONAL DIAGNOSIS

Based on clinical presentations, FBE results, and morphology, this case is highly suggestive of hereditary stomatocytosis and further tests are suggested.

DIFFERENTIAL DIAGNOSES

Acute alcoholic intoxication
Severe hepatobiliary disease
Certain drugs
Other inherited condition
Rh—deficiency disease

FURTHER TESTS AND EXPECTED RESULTS

Confirmation of Provisional Diagnosis

Osmotic fragility → increased
Biochemistry → bilirubinaemia, haptoglobinaemia, and ferritinaemia
Molecular tests → defect in band 7.2, 4.2, and band 3

Exclusion of Differential Diagnoses

Blood grouping → to exclude Rh null disease

FINAL DIAGNOSIS

Hereditary stomatocytosis is the final diagnosis for this case based on further tests that confirm the provisional diagnosis and exclude all the differential diagnoses.

PATHOPHYSIOLOGY OF HEREDITARY STOMATOCYTOSIS

Hereditary stomatocytosis is characterized morphologically by stomatocytes and biochemically by the failure of the Na+ and K+ and increased in intracellular water. The influx of Na+ exceeds the loss of K+, and the cells swell, becoming less dense and more stomatocytic. In most patients, a deficiency of a membrane integral protein called stomatin or band 7.2b occurs [2].

Case 21

Clinical presentation

● Mild features of anaemia

FBE RESULTS

CBC (Complete Blood Count) Parameters

Parameters	Results	Units
Hb	110	g/L
RCC	3.60	$\times 10^{12}\,L^{-1}$
PCV	0.36	
MCV	100	Fl
MCH	31.0	Pg
MCHC	306	g/L
RDW	19.8	%
Retics	N/A	%
WBC	8.2	$\times 10^9\,L^{-1}$
Plt count	255	$\times 10^9\,L^{-1}$

Differential White Cell Count

Cell Type	Percentages (%)	Absolute Count
Neutrophils	68	$5.58 \times 10^9\,L^{-1}$
Bands	0	$0 \times 10^9\,L^{-1}$
Lymphocytes	23	$1.89 \times 10^9\,L^{-1}$
Monocytes	7	$0.57 \times 10^9\,L^{-1}$
Eosinophils	1	$0.08 \times 10^9\,L^{-1}$
Basophils	1	$0.08 \times 10^9\,L^{-1}$
Metamyelocytes	0	
Myelocytes	0	

Continued

Differential White Cell Count—cont'd		
Cell Type	Percentages (%)	Absolute Count
Promyelocytes	0	
Blasts	0	
Prolymphocytes	0	
nRBC/100WBC	0	

RBC MORPHOLOGY

Marked anisocytosis, mild macrocytes, mild ovalocytes, moderate target cells, and mild stomatocytes.

WBC MORPHOLOGY

Normal in number and morphology.

PLATELET MORPHOLOGY

Normal in number and morphology.

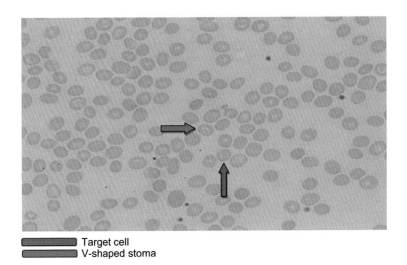

▭ Target cell
▭ V-shaped stoma

Blood film image shows the most diagnostic features.

PROVISIONAL DIAGNOSIS

Based on clinical presentations, FBE results, and morphology, this case is highly suggestive of Melanesian ovalocytosis and further tests are suggested.

DIFFERENTIAL DIAGNOSES

This case is known from the blood film and it is unlikely to be confused with other condition.

FURTHER TESTS AND EXPECTED RESULTS

Confirmation of Provisional Diagnosis

Osmotic fragility → decreased
Thermal stability → increased
Molecular tests → reduced expression of many red cell antigens and defect in band 3
Flow cytometry → abnormal binding of eosin-5-maleimide

FINAL DIAGNOSIS

Melanesian ovalocytosis is the final diagnosis for this case based on further tests that confirm the provisional diagnosis and exclude all the differential diagnoses.

PATHOPHYSIOLOGY

Hereditary Melanesian ovalocytosis, also sometimes referred to as southeast Asian ovalocytosis is a distinct and homogenous disorder that occurs in Melanesians in Papua and New Guinea and in Malaysian aboriginals and the population of Indonesia and the Philippines. The underlying genetic defect is deletion of nine codons in the SLC4A1 gene for band 3 which results in tight binding of band 3 to ankyrin, reduced lateral mobility, and rigidity of the membrane, leading to the formation of ovalocytes [3].

The membrane rigidity acts as a protective mechanism against all strains of malaria, including resistance to cerebral malaria caused by *Plasmodium falciparum*.

Case 22

Clinical presentations

- General features of anaemia
- Growth retardation
- Splenomegaly

FBE RESULTS

CBC (Complete Blood Count) Parameters

Parameters	Results	Units
Hb	87	g/L
RCC	3.55	$\times 10^{12}\,L^{-1}$
PCV	0.24	
MCV	68	Fl
MCH	24.5	Pg
MCHC	363	g/L
RDW	26.8	%
Retics	N/A	%
WBC	8.5	$\times 10^{9}\,L^{-1}$
Plt count	150	$\times 10^{9}\,L^{-1}$

Differential White Cell Count

Cell Type	Percentages (%)	Absolute count
Neutrophils	65	$5.52 \times 10^{9}\,L^{-1}$
Bands	0	$0 \times 10^{9}\,L^{-1}$
Lymphocytes	22	$1.87 \times 10^{9}\,L^{-1}$
Monocytes	7	$0.59 \times 10^{9}\,L^{-1}$
Eosinophils	3	$0.25 \times 10^{9}\,L^{-1}$
Basophils	3	$0.25 \times 10^{9}\,L^{-1}$
Metamyelocytes	0	

Differential White Cell Count—cont'd

Cell Type	Percentages (%)	Absolute count
Myelocytes	0	
Promyelocytes	0	
Blasts	0	
Prolymphocytes	0	
nRBC/100WBC	0	

RBC MORPHOLOGY

Marked anisocytosis, mild polychromasia, marked poikilocytosis, moderate crenated cells, and mild spherocytes.

WBC MORPHOLOGY

Normal in number and morphology.

PLATELET MORPHOLOGY

Normal in number and morphology.

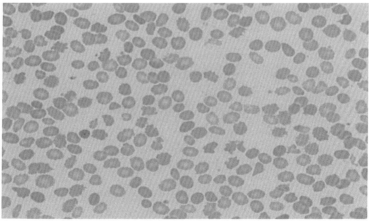

The most diagnostic features is marked poikilocytosis

Blood film image shows the most diagnostic features.

PROVISIONAL DIAGNOSIS

Based on clinical presentations, FBE results, and morphology, this case is highly suggestive of hereditary pyropoikilocytosis and further tests are suggested.

DIFFERENTIAL DIAGNOSES

Hereditary elliptocytosis
Haemoglobin H disease
Congenital dyserythropoietic anaemia
Severe burns

FURTHER TESTS AND EXPECTED RESULTS

Confirmation of Provisional Diagnosis

Reticulocyte count → increased
DAT → negative
Osmotic fragility → increased
Biochemical analysis → demonstration of fragmentation
Flow cytometry → abnormal binding of eosin-5-maleimide
Thermal sensitivity → the spectrin denatures at temperatures between 45°C and 46°C (normal 49–50)
Molecular tests → defect in alpha spectrin

FINAL DIAGNOSIS

Hereditary pyropoikilocytosis is the final diagnosis for this case based on further tests that confirm the provisional diagnosis and exclude all the differential diagnoses.

PATHOPHYSIOLOGY OF HEREDITARY PYROPOIKILOCYTOSIS

Hereditary pyropoikilocytosis is a heterogeneous group of inherited haemolytic anaemias characterized by recessive inheritance and bizarre poikilocytes including red cell fragments and microspherocytes. It has been described in Caucasian, Black, and Arab populations. The condition is defined by enhanced red cell fragmentation on in vitro heating which occurs at a lower temperature than with normal red cells. This feature is indicated in the name 'pyropoikilocytosis'. Red cell membranes often show two defects, a partial spectrin deficiency and a defect in the self-assembly of spectrin dimers into tetramers, the latter as a result of elliptogenic mutation. It may be the spectrin deficiency that leads to the presence of spherocytes as well as elliptocytes in patients with hereditary pyropoikilocytosis. The underlying genetic defects are various.

There may be homozygosity or compound heterozygosity for a mutant spectrin that has a defect affecting dimer self-assembly and is also degraded rapidly. Alternatively, there may be compound heterozygosity for a mutant spectrin (alpha or beta chain) and for a defect leading to a reduced rate of synthesis of alpha spectrin, alpha spectrin solely occurring in trans. Parents of patients with hereditary pyropoikilocytosis may both have morphologically normal red cells or one or occasionally both parents may have typical hereditary elliptocytosis [3].

Case 23

Clinical presentations

- Weakness, fatigue, short of breath, and back pain
- Trouble sleeping and haemoglobinuria on voiding after sleep
- Jaundice

FBE RESULTS

CBC (Complete Blood Count) Parameters

Parameters	Results	Units
Hb	101	g/L
RCC	3.40	$\times 10^{12}\,L^{-1}$
PCV	0.37	
MCV	109	Fl
MCH	30.0	Pg
MCHC	273	g/L
RDW	17.4	%
Retics	N/A	%
WBC	4.5	$\times 10^{9}\,L^{-1}$
Plt count	65	$\times 10^{9}\,L^{-1}$

Differential White Cell Count

Cell Type	Percentages (%)	Absolute Count
Neutrophils	65	$2.92 \times 10^{9}\,L^{-1}$
Bands	4	$0.18 \times 10^{9}\,L^{-1}$
Lymphocytes	26	$1.17 \times 10^{9}\,L^{-1}$
Monocytes	4	$0.18 \times 10^{9}\,L^{-1}$
Eosinophils	1	$0.04 \times 10^{9}\,L^{-1}$
Basophils	0	$0 \times 10^{9}\,L^{-1}$
Metamyelocytes	0	

Differential White Cell Count—cont'd

Cell Type	Percentages (%)	Absolute Count
Myelocytes	0	
Promyelocytes	0	
Blasts	0	
Prolymphocytes	0	
nRBC/100WBC	0	

RBC MORPHOLOGY

Moderate anisocytosis, moderate macrocytes, mild polychromasia, mild target cells, mild elongated and mild irregular contracted cells.

WBC MORPHOLOGY

Normal in number and morphology.

PLATELET MORPHOLOGY

Moderate thrombocytopenia with normal morphology.

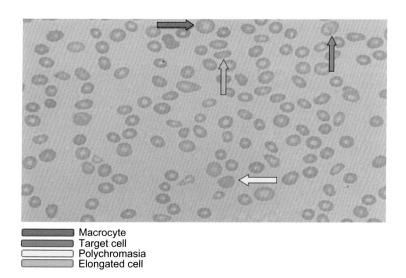

◼ Macrocyte
◼ Target cell
▢ Polychromasia
◼ Elongated cell

Blood film image shows the most diagnostic features.

PROVISIONAL DIAGNOSIS

Based on clinical presentations, FBE results, and morphology, this case is highly suggestive of paroxysmal nocturnal haemoglobinuria (PNH) and further tests are suggested.

DIFFERENTIAL DIAGNOSES

Paroxysmal cold haemoglobinuria
Other haemolytic anaemia
Aplastic anaemia
Myelodysplastic syndrome
Renal disease

FURTHER TESTS AND EXPECTED RESULTS

Confirmation of Provisional Diagnosis

Chemistries

- **Sugar lysis test** → positive
- **Ham test** → positive
- **Complement lysis sensitivity test** → positive in type I and II
- **LDH** → elevated
- **Haptoglobin** → low to absent
- **Urinalysis** → Haemoglobinuria and haemosiderinuria (using Perls' Prussian blue)

    Flow cytometry → deficient or absent CD55 and CD59
    Genetic → mutation in phosphatidylinositol-glycan complementation class A (PIG-A) gene
    Bone marrow → marrow hypoplasia

Exclusion of Differential Diagnoses

    Osmotic fragility test → negative
    Any abnormal cytogenetic associated with myelodysplasia → negative exclude MDS

FINAL DIAGNOSIS

Paroxysmal nocturnal haemoglobinuria (PNH) is the final diagnosis for this case based on further tests that confirm the provisional diagnosis and exclude all the differential diagnoses.

PATHOPHYSIOLOGY OF PNH

Paroxysmal nocturnal haemoglobinuria is caused by an acquired clonal stem mutation in the PGIA gene that leads to the production of RBCs' that lack glycosyl-phosphatidylinositol (GPI) membrane proteins. The absence of GPI-linked proteins increases the susceptibility of complement binding to RBCs'. In PNH two normal complement inhibitor proteins, decay accelerating factor (DAF, CD55) and membrane inhibitor of reactive lysis (MIRL, CD59), were shown to be absent. The CD55 protein is known to inhibit the formation of C3 and C3 convertases of the alternative complement system, while, CD59 proteins prevent the formation of membrane attack complex. An absence of these proteins will consequently lead to incidental complement and C3 convertase complex formation on the cell membrane, thus leading to formation of the membrane attack complex, membrane pores, and cell lysis. Since PNH is associated with mutations in stem cells, the absence of GPI-linked membrane proteins can also be seen in platelets and WBC's. In severe cases patients with PNH may experience bone marrow dysfunction leading to marked anaemia [2].

Chapter 7

Immune Haemolytic Disorders

Case 24

Clinical presentations

- Fatigue
- Pale skin colour
- Rapid heart rate
- Shortness of breath

FBE RESULTS

CBC (Complete Blood Count) Parameters

Parameters	Results	Units
Hb	90	g/L
RCC	2.57	$\times 10^{12} \, L^{-1}$
PCV	0.28	
MCV	109	Fl
MCH	35.0	Pg
MCHC	321	g/L
RDW	23.2	%
Retics	N/A	%
WBC	20.0	$\times 10^9 \, L^{-1}$
Plt count	85	$\times 10^9 \, L^{-1}$

Differential White Cell Count

Cell Type	Percentages (%)	Absolute Count
Neutrophils	69	$13.8 \times 10^9 \, L^{-1}$
Bands	5	$1.0 \times 10^9 \, L^{-1}$
Lymphocytes	10	$2.0 \times 10^9 \, L^{-1}$
Monocytes	6	$1.2 \times 10^9 \, L^{-1}$
Eosinophils	1	$0.2 \times 10^9 \, L^{-1}$
Basophils	0	$0 \times 10^9 \, L^{-1}$
Metamyelocytes	4	
Myelocytes	5	
Promyelocytes	0	
Blasts	0	
Prolymphocytes	0	
nRBC/100WBC	2	

RBC MORPHOLOGY

Marked anisocytosis, moderate polychromasia, moderate macrocytes, moderate bite cells, occasional spherocytes, occasional target cells, occasional fragment cells, occasional Howell–Jolly bodies (HJB), and occasional nRBC.

WBC MORPHOLOGY

Mild leukococytosis, moderate neutrophilia, and mild left shift.

PLATELET MORPHOLOGY

Mild thrombocytopenia with normal morphology.

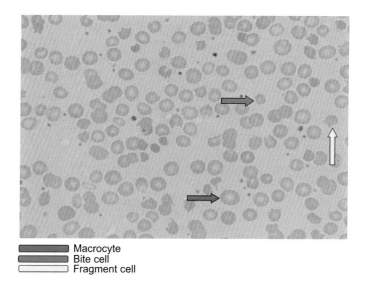

☐ Macrocyte
☐ Bite cell
☐ Fragment cell

Blood film image shows the most diagnostic features.

PROVISIONAL DIAGNOSIS

Based on clinical presentations, FBE results, and morphology, this case is highly suggestive of drug-induced haemolytic anaemia and further tests are suggested.

DIFFERENTIAL DIAGNOSES

Methemoglobinaemia
Autoimmune haemolytic anaemia
Other haemolytic anaemia such as glucose-6-phosphate dehydrogenase deficiency

FURTHER TESTS AND EXPECTED RESULTS

- **Direct Coombs' test** → positive.
- **Indirect bilirubin** → high.
- **Serum haptoglobin** → low.
- **Urine analysis** → haemoglobinuria and haemosiderinuria
- **Urine and faecal urobilinogen** → increased.
- **History of drugs** → positive.

FINAL DIAGNOSIS

Drug-induced haemolytic anaemia is the final diagnosis for this case based on further tests that confirm the provisional diagnosis and exclude all the differential diagnoses.

PATHOPHYSIOLOGY OF DRUG-INDUCED HAEMOLYTIC ANAEMIA (DIHA)

Drugs may cause immune haemolytic anaemia via three mechanisms:

1. Antibody directed against a drug–red cell membrane complex (e.g. penicillin, ampicillin).
2. Deposition of complement via a drug (protein)–antibody complex onto the red cell surface (e.g. quinidine, rifampicin).
3. A true autoimmune haemolytic anaemia in which the role of the drug is unclear (e.g. methyldopa).

In each case, the haemolytic anaemia gradually disappears when the drug is discontinued but with methyldopa the autoantibody may persist for several months. The penicillin-induced immune haemolytic anaemia only occurs with massive doses of the antibiotic [12].

Case 25

Clinical presentations

- Tired
- Short of breath
- Pallor

FBE RESULTS

CBC (Complete Blood Count) Parameters

Parameters	Results	Units
Hb	55	g/L
RCC	1.60	$\times 10^{12}$ L^{-1}
PCV	0.14	
MCV	88	Fl
MCH	34.0	Pg
MCHC	393	g/L
RDW	18.0	%
Retics	36.0	%
WBC	10.0	$\times 10^{9}$ L^{-1}
Plt count	439	$\times 10^{9}$ L^{-1}

Differential White Cell Count

Cell Type	Percentages (%)	Absolute Count
Neutrophils	83	8.3×10^{9} L^{-1}
Bands	0	0×10^{9} L^{-1}
Lymphocytes	10	1.0×10^{9} L^{-1}
Monocytes	5	0.5×10^{9} L^{-1}
Eosinophils	2	0.2×10^{9} L^{-1}
Basophils	0	0×10^{9} L^{-1}
Metamyelocytes	0	

Continued

Differential White Cell Count—cont'd		
Cell Type	Percentages (%)	Absolute Count
Myelocytes	0	
Promyelocytes	0	
Blasts	0	
Prolymphocytes	0	
nRBC/100WBC	8	

RBC MORPHOLOGY

Moderate anisocytosis, marked polychromasia, marked microspherocytes, occasional HJB, and occasional Pappenheimer cells.

WBC MORPHOLOGY

Normal in number and morphology.

PLATELET MORPHOLOGY

Mild thrombocytosis with normal morphology.

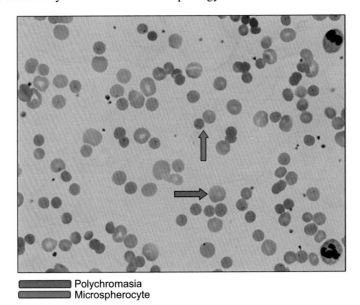

▬ Polychromasia
▬ Microspherocyte

Blood film image shows the most diagnostic features.

PROVISIONAL DIAGNOSIS

Based on clinical presentations, FBE results, and morphology, this case is highly suggestive of warm autoimmune haemolytic anaemia and further tests are suggested.

DIFFERENTIAL DIAGNOSES

Cold autoimmune haemolytic anaemia (AIHA)
Drug-induced haemolytic anaemia (DIHA)
Idiopathic haemolytic anaemia

FURTHER TESTS AND EXPECTED RESULTS

DAT → positive with anti-IgG and negative with complement C3d

FINAL DIAGNOSIS

Warm autoimmune haemolytic anaemia is the final diagnosis for this case based on further tests that confirm the provisional diagnosis and exclude all the differential diagnoses.

PATHOPHYSIOLOGY OF WARM AUTOIMMUNE HAEMOLYTIC ANAEMIA (wAIHA)

Autoimmune haemolytic anaemias (AIHAs) are caused by antibodies produced by the body against its own red cells. In AIHA the red cells are coated with antibodies, usually immunoglobulin G (IgG) alone or with complement and are therefore taken up by RE macrophages which have receptors for the Ig Fc fragment. As a result, part of the coated membrane is lost, therefore the cell becomes progressively more spherical to maintain the same volume and is ultimately prematurely destroyed, predominantly in the spleen. When the cells are coated with IgG or complement (C3d, the degraded fragment of C3), red cell destruction occurs more generally in the RE system. In wAIHA, the antibody reacts more strongly with red cells at 37°C [12].

Case 26

Clinical presentations

- Tired
- Short of breath
- Pallor

FBE RESULTS

CBC (Complete Blood Count) Parameters

Parameters	Results	Units
Hb	84	g/L
RCC	2.71	$\times 10^{12}\, L^{-1}$
PCV	0.22	
MCV	81	Fl
MCH	31.0	Pg
MCHC	382	g/L
RDW	21.9	%
Retics	N/A	%
WBC	5.5	$\times 10^{9}\, L^{-1}$
Plt count	85	$\times 10^{9}\, L^{-1}$

Differential White Cell Count

Cell Type	Percentages (%)	Absolute Count
Neutrophils	47	$2.58 \times 10^{9}\, L^{-1}$
Bands	14	$0.77 \times 10^{9}\, L^{-1}$
Lymphocytes	29	$1.59 \times 10^{9}\, L^{-1}$
Monocytes	12	$0.66 \times 10^{9}\, L^{-1}$
Eosinophils	0	$0 \times 10^{9}\, L^{-1}$
Basophils	0	$0 \times 10^{9}\, L^{-1}$
Metamyelocytes	0	

Differential White Cell Count—cont'd

Cell Type	Percentages (%)	Absolute Count
Myelocytes	1	
Promyelocytes	0	
Blasts	0	
Prolymphocytes	0	
nRBC/100WBC	6	

RBC MORPHOLOGY

Marked anisocytosis, marked polychromasia, occasional macrocytes, marked irregular contracted cells, marked crenated cells, and moderate agglutination.

WBC MORPHOLOGY

Normal in number with mild left shift and occasional atypical lymphocyte.

PLATELET MORPHOLOGY

Moderate thrombocytopenia with normal morphology.

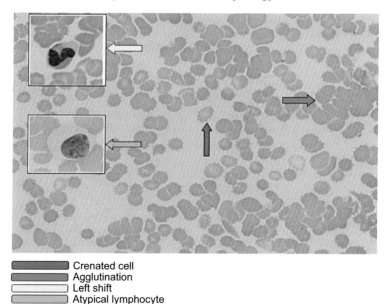

▬ Crenated cell
▬ Agglutination
▭ Left shift
▬ Atypical lymphocyte

Blood film image shows the most diagnostic features.

PROVISIONAL DIAGNOSIS

Based on clinical presentations, FBE results, and morphology, this case is highly suggestive of cold autoimmune haemolytic anaemia and further tests are suggested.

DIFFERENTIAL DIAGNOSES

Warm autoimmune haemolytic anaemia (AIHA)
Drug-induced haemolytic anaemia (DIHA)
Idiopathic haemolytic anaemia

FURTHER TESTS AND EXPECTED RESULTS

DAT → positive with complement C3d and negative with anti-IgG

FINAL DIAGNOSIS

Cold autoimmune haemolytic anaemia is the final diagnosis for this case based on further tests that confirm the provisional diagnosis and exclude all the differential diagnoses.

PATHOPHYSIOLOGY OF COLD AUTOIMMUNE HAEMOLYTIC ANAEMIA (cAIHA)

In this syndrome the autoantibody, whether monoclonal (as in the idiopathic cold haemagglutinin syndrome or associated with lymphoproliferative disorders) or polyclonal (as following infection, e.g. infectious mononucleosis) attaches to red cells mainly in the peripheral circulation where the temperature of blood can drop significanlty. cAIHA is usually associated with the production of immunoglobulin M (IgM) that reacts optimally with red cells at 4°C. IgM antibodies are highly efficient at fixing complement on red cells and therefore, both intravascular and extravascular haemolysis can occur in the patient. Usually only complement proteins are detected on the red cells, as the antibody would be eluted off the cells in warmer parts of the circulation [12].

Case 27

CLINICAL PRESENTATIONS

Newborn baby has

- General features of anaemia
- Jaundice
- Enlarged liver and spleen

FBE RESULTS

CBC (Complete Blood Count) Parameters

Parameters	Results	Units
Hb	63	g/L
RCC	1.70	$\times 10^{12}\,L^{-1}$
PCV	0.20	
MCV	118	Fl
MCH	37.1	Pg
MCHC	315	g/L
RDW	22.3	%
Retics	N/A	%
WBC	12.2 Corrected to 10.7	$\times 10^{9}\,L^{-1}$
Plt count	200	$\times 10^{9}\,L^{-1}$

Differential White Cell Count

Cell Type	Percentages (%)	Absolute Count
Neutrophils	29	$3.10 \times 10^{9}\,L^{-1}$
Bands	10	$1.07 \times 10^{9}\,L^{-1}$
Lymphocytes	38	$4.07 \times 10^{9}\,L^{-1}$
Monocytes	14	$1.50 \times 10^{9}\,L^{-1}$
Eosinophils	0	$0 \times 10^{9}\,L^{-1}$

Continued

Differential White Cell Count—cont'd		
Cell Type	Percentages (%)	Absolute Count
Basophils	1	$0.11 \times 10^9 \, L^{-1}$
Metamyelocytes	0	
Myelocytes	8	
Promyelocytes	0	
Blasts	0	
Prolymphocytes	0	
nRBC/100WBC	14	

RBC MORPHOLOGY

Moderate anisocytosis, moderate macrocytes, marked polychromasia, moderate spherocytes, marked crenated cells, and moderate nRBC.

WBC MORPHOLOGY

Normal in number and morphology.

PLATELET MORPHOLOGY

Normal in number and morphology.

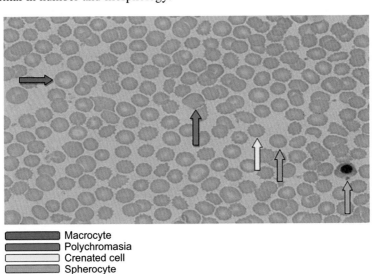

◼ Macrocyte
◼ Polychromasia
◻ Crenated cell
◻ Spherocyte
◻ nRBC

Blood film image shows the most diagnostic features.

PROVISIONAL DIAGNOSIS

Based on clinical presentations, FBE results, and morphology, this case is highly suggestive of Rh haemolytic disease of the newborn (Rh-HDNB) and further tests are suggested.

DIFFERENTIAL DIAGNOSIS

Hereditary spherocytosis

FURTHER TESTS AND EXPECTED RESULTS

Confirmation of Provisional Diagnosis

DAT → positive (baby) and detection of IgG antibody in maternal serum against the foetal red cell antigen.

Exclusion of Differential Diagnosis

Osmotic fragility test → negative
Acidified glycerol lysis test (AGLT) → negative
Pink test → negative

FINAL DIAGNOSIS

Rh haemolytic disease of the newborn (Rh-HDNB) is the final diagnosis for this case based on further tests that confirm the provisional diagnosis and exclude all the differential diagnoses.

PATHOPHYSIOLOGY OF Rh-HDNB

When an Rh D-negative woman is pregnant with an Rh D-positive foetus, the Rh D-positive foetal red cells cross into the maternal circulation (especially at parturition and during the third trimester) and sensitize the mother to form anti-D antibodies. The mother could also be sensitized by a previous miscarriage, amniocentesis, or other trauma to the placenta or by blood transfusion. Anti-D antibodies formed in the mother can cross the placenta and coat the foetal red cells during the next pregnancy with an Rh D-positive foetus, and causes reticuloendothelial destruction of these cells leading to anaemia and jaundice in the foetus. If the father is heterozygous for D antigen, there is a 50% probability that the foetus will be D-positive. The main aim of management is to prevent anti-D antibody formation in Rh D-negative mothers. This can be achieved by the administration of small amounts of anti-D antibody which 'mop up' and destroy Rh D-positive foetal red cells in the mother before they can sensitize the immune system of the mother to produce anti-D [12].

Chapter 8

Acute Leukaemias

Case 28

Clinical presentations

- Fever
- Fatigue
- Weight loss
- Easy bruising
- Hepatosplenomegaly

FBE RESULTS

CBC (Complete Blood Count) Parameters		
Parameters	Results	Units
Hb	98	g/L
RCC	3.30	$\times 10^{12}\,L^{-1}$
PCV	0.34	
MCV	103	Fl
MCH	30.0	Pg
MCHC	288	g/L
RDW	17.0	%
Retics	N/A	%
WBC	13.8	$\times 10^{9}\,L^{-1}$
Plt count	50	$\times 10^{9}\,L^{-1}$

Differential White Cell Count

Cell Type	Percentages (%)	Absolute Count
Neutrophils	4	$0.55 \times 10^9 \, L^{-1}$
Bands	0	$0 \times 10^9 \, L^{-1}$
Lymphocytes	8	$1.1 \times 10^9 \, L^{-1}$
Monocytes	0	$0 \times 10^9 \, L^{-1}$
Eosinophils	0	$0 \times 10^9 \, L^{-1}$
Basophils	0	$0 \times 10^9 \, L^{-1}$
Metamyelocytes	0	
Myelocytes	0	
Promyelocytes	0	
Blasts	88	
Prolymphocytes	0	
nRBC/100WBC	0	

RBC MORPHOLOGY

Moderate anisocytosis, mild polychromasia, mild macrocytes, and mild teardrops.

WBC MORPHOLOGY

Mild leucocytosis, marked neutropenia, many medium-sized blasts which have fine chromatin pattern, small nucleoli (more than two per cell), and high nucleocytoplasmic ratio.

PLATELET MORPHOLOGY

Moderate thrombocytopenia with normal morphology.

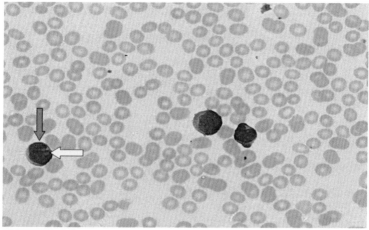

It is very hard to distinguish blasts from a blood film; however, there are characteristics for lymphoblasts and myeloblasts that can distinguish them from each other such as in this film, the myeloblast → medium-sized blast which has a fine chromatin pattern, (small nucleoli ⬚), and high nucleocyto plasmic ratio.

Blood film image shows the most diagnostic features.

PROVISIONAL DIAGNOSIS

Based on clinical presentations, FBE results, and morphology, this case is highly suggestive of acute myeloid leukaemia without maturation and further tests are suggested.

DIFFERENTIAL DIAGNOSES

ALL (acute lymphoblastic leukaemia)
APML (acute promyelocytic leukaemia)
AMML (acute myelomonocytic leukaemia)
Leukemoid reaction
MDS transformation (myelodysplastic syndrome)
Aplastic anaemia

FURTHER TESTS AND EXPECTED RESULTS

Confirmation of Provisional Diagnosis

Bone marrow → more than 90% blast cells and less than 10% promyelo-cytes and markedly hypercellular.

Cytochemistry

- **Myeloperoxidase (MPO)** → positive
- **Sudan black B (SBB)** → positive
- **Chloroacetate esterase (CAE)** → positive

    Immunophenotyping → CD13, 33, 117, & 34 (myeloid markers) are positive.

Exclusion of Differential Diagnoses

Cytochemistry

- **Periodic Acid Schiff (PAS)** → negative
- **Acid phosphatase** → negative
- **Alpha naphthyl acetate esterase (ANAE)** → negative

Immunophenotyping →

    B (CD10, 22) & T (CD 2, 3, 5, 7) cell markers are negative
    CD14&64 → negative

    Cytogenetic and molecular study → t (15:17) & PML-RAR should be negative to exclude APML.

FINAL DIAGNOSIS

Acute myeloid leukaemia without maturation is the final diagnosis for this case based on further tests that confirm the provisional diagnosis and exclude all the differential diagnoses.

Case 29

Clinical presentations

- Fever
- Fatigue
- Weight loss
- Easy bruising
- Hepatosplenomegaly

FBE RESULTS

CBC (Complete Blood Count) Parameters

Parameters	Results	Units
Hb	90	g/L
RCC	3.16	$\times 10^{12}\,L^{-1}$
PCV	0.26	
MCV	82	Fl
MCH	28.5	Pg
MCHC	346	g/L
RDW	20.0	%
Retics	N/A	%
WBC	55.0	$\times 10^9\,L^{-1}$
Plt count	25	$\times 10^9\,L^{-1}$

Differential White Cell Count

Cell Type	Percentages (%)	Absolute Count
Neutrophils	12	$6.6 \times 10^9\,L^{-1}$
Bands	13	$7.15 \times 10^9\,L^{-1}$
Lymphocytes	6	$3.3 \times 10^9\,L^{-1}$
Monocytes	0	$0 \times 10^9\,L^{-1}$
Eosinophils	1	$0.55 \times 10^9\,L^{-1}$

Continued

Differential White Cell Count—cont'd		
Cell Type	Percentages (%)	Absolute Count
Basophils	0	$0 \times 10^9 \, L^{-1}$
Metamyelocytes	0	
Myelocytes	10	
Promyelocytes	10	
Blasts	48	
Prolymphocytes	0	
nRBC/100WBC	0	

RBC MORPHOLOGY

Moderate anisocytosis, mild polychromasia, mild spherocytes, and mild irregular contracted cells.

WBC MORPHOLOGY

Marked leucocytosis with many blasts which are large, high N/C ratio, 1–3 prominent nucleoli, fine chromatin, and occasional Auer rods.

PLATELET MORPHOLOGY

Marked thrombocytopenia with normal morphology.

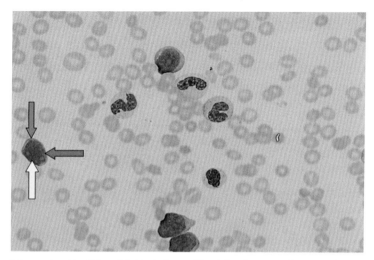

It is very hard to distinguish blasts from a blood film, however, there are characteristics for lymphoblasts and myeloblasts that can distinguish them from each other such as in this film, the myeloblast → medium-sized blast which has a fine chromatin pattern, (small nucleoli [＿＿＿＿＿]), and high nucleocytoplasmic ratio.

[▬▬▬▬▬] Nucleus (round—oval, 7:1 to 4:1 ratio, reddish to purple colour and up to 5 nucleoli)

[▬▬▬▬▬] Cytoplasm (pale—deep blue, lighter near nucleus and usually no granules)

Blood film image shows the most diagnostic features.

PROVISIONAL DIAGNOSIS

Based on clinical presentations, FBE results, and morphology, this case is highly suggestive of acute myeloid leukaemia with t(8:21)(q22:q22)-AML1/ETO and further tests are suggested.

DIFFERENTIAL DIAGNOSES

ALL (acute lymphoblastic leukaemia)
APML (acute promyelocytic leukaemia)
AMML (acute myelomonocytic leukaemia)
Leukemoid reaction
MDS transformation (myelodysplastic syndrome)
Aplastic anaemia

FURTHER TESTS AND EXPECTED RESULTS

Confirmation of Provisional Diagnosis

Bone marrow → 30%–89% blast cells and more than 10% promyelocytes or neutrophils (precursors) and hypercellular.

Cytochemistry

- **Myeloperoxidase (MPO)** → positive
- **Sudan black B (SBB)** → positive
- **Chloroacetate esterase (CAE)** → positive

Immunophenotyping → CD13,33,117 & 34 (myeloid markers) are positive.
Cytogenetic and molecular tests → t(8:21)-AML1/ETO

Exclusion of Differential Diagnoses

Cytochemistry

- **Periodic Acid Schiff (PAS)** → negative
- **Acid phosphatase** → negative
- **Alpha naphthyl acetate esterase (ANAE)** → negative

Immunophenotyping → B cells (CD10 & 22) & T cells (CD 2, 3, 5, & 7) cell markers are negative, CD(14 & 64) → negative
Cytogenetic and molecular study → t(15:17) & PML-RAR should be negative to exclude APML.

FINAL DIAGNOSIS

Acute myeloid leukaemia with t(8:21)(q22:q22)-AML1 is the final diagnosis for this case based on further tests that confirm the provisional diagnosis and exclude all the differential diagnoses.

Case 30

Clinical presentations

- Fever
- Fatigue
- Weight loss
- Easy bruising
- Hepatosplenomegaly

FBE RESULTS

CBC (Complete Blood Count) Parameters

Parameters	Results	Units
Hb	134	g/L
RCC	4.10	$\times 10^{12} \, L^{-1}$
PCV	0.39	
MCV	93	Fl
MCH	32.0	Pg
MCHC	347	g/L
RDW	13.9	%
Retics	N/A	%
WBC	26.3	$\times 10^{9} \, L^{-1}$
Plt count	41	$\times 10^{9} \, L^{-1}$

Differential White Cell Count

Cell Type	Percentages (%)	Absolute Count
Neutrophils	10	$2.63 \times 10^{9} \, L^{-1}$
Bands	0	$0 \times 10^{9} \, L^{-1}$
Lymphocytes	17	$4.47 \times 10^{9} \, L^{-1}$
Monocytes	3	$0.79 \times 10^{9} \, L^{-1}$
Eosinophils	0	$0 \times 10^{9} \, L^{-1}$
Basophils	0	$0 \times 10^{9} \, L^{-1}$
Metamyelocytes	0	

Continued

Differential White Cell Count—cont'd		
Cell Type	Percentages (%)	Absolute Count
Myelocytes	0	
Promyelocytes	28	
Blasts	42	
Prolymphocytes	0	
nRBC/100WBC	0	

RBC MORPHOLOGY

Normocytic cells, mild polychromasia, and occasional nRBC.

WBC MORPHOLOGY

Marked leucocytosis, marked left shift, moderate Auer rods, and many abnormal promyelocytes with fine azurophilic granules and multiple Auer rods. Myeloblast with occasional Auer rods present.

PLATELET MORPHOLOGY

Marked thrombocytopenia with normal morphology.

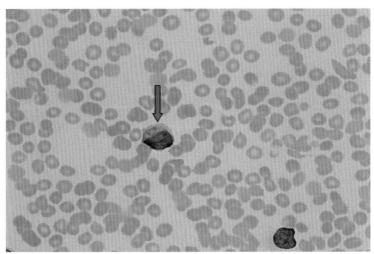

Abnormal promyelocytes with fine azurophilic granules and multiple auer rods.

Blood film image shows the most diagnostic features.

PROVISIONAL DIAGNOSIS

Based on clinical presentations, FBE results, and morphology, this case is highly suggestive of acute promyelocytic leukaemia with t(15:17)-PML-RAR and further tests are suggested.

DIFFERENTIAL DIAGNOSES

AML (acute myeloid leukaemia)
APML-DIC
ALL (acute lymphoblastic leukaemia)
AMML (acute myelomonocytic leukaemia)
Leukemoid reaction
MDS transformation (myelodysplastic syndrome)
Aplastic anaemia

FURTHER TESTS AND EXPECTED RESULTS

Confirmation of Provisional Diagnosis

Bone marrow → more than 30% blast cells.

Cytochemistry

- **Myeloperoxidase (MPO)** → positive
- **Sudan black B (SBB)** → positive
- **Chloroacetate esterase (CAE)** → positive

   Immunophenotyping → CD13 & 33 (myeloid markers) are positive, coexpression of CD2 & 9.
   Cytogenetic → t(15:17)
   Molecular studies → PML-RAR gene

Exclusion of Differential Diagnoses

Cytochemistry

- **Periodic Acid Schiff (PAS)** → negative
- **Acid phosphatase** → negative
- **Alpha naphthyl acetate esterase (ANAE)** → negative

Immunophenotyping →

   B (CD10, 19, 22) & T (CD 2, 3, 5, 7) cell markers are negative
   CD14 & 64 → negative

   Coagulation tests → normal to rule out DIC.

FINAL DIAGNOSIS

Acute promyelocytic leukaemia with t(15:17)-PML-RARA is the final diagnosis for this case based on further tests that confirm the provisional diagnosis and exclude all the differential diagnoses.

Case 31

Clinical presentations

- Fever
- Pallor
- Easy bruising
- Neurological symptoms

FBE RESULTS

CBC (Complete Blood Count) Parameters

Parameters	Results	Units
Hb	82	g/L
RCC	3.10	$\times 10^{12}\,L^{-1}$
PCV	0.29	
MCV	94	Fl
MCH	26.0	Pg
MCHC	283	g/L
RDW	16.9	%
Retics	N/A	%
WBC	6.0	$\times 10^9\,L^{-1}$
Plt count	4	$\times 10^9\,L^{-1}$

Differential White Cell Count

Cell Type	Percentages (%)	Absolute Count
Neutrophils	18	$1.08 \times 10^9\,L^{-1}$
Bands	0	$0 \times 10^9\,L^{-1}$
Lymphocytes	9	$0.54 \times 10^9\,L^{-1}$
Monocytes	26	$1.56 \times 10^9\,L^{-1}$
Eosinophils	0	$0 \times 10^9\,L^{-1}$
Basophils	0	$0 \times 10^9\,L^{-1}$
Metamyelocytes	0	

Continued

Differential White Cell Count—cont'd		
Cell Type	Percentages (%)	Absolute Count
Myelocytes	0	
Promyelocytes	0	
Blasts	47	
Prolymphocytes	0	
nRBC/100WBC	0	

RBC MORPHOLOGY

Moderate anisocytosis and marked rouleaux.

WBC MORPHOLOGY

Large blasts with prominent nucleoli, many have Auer rods, some show monoblastic features.

PLATELET MORPHOLOGY

Marked thrombocytopenia with normal morphology

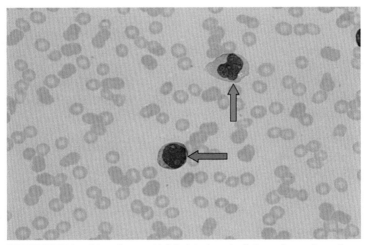

Monoblast (large with lobulated nucleoli, lacy chromatin pattern, several nucleoli, and voluminous finely granulated cytoplasm)
Myeloblast (small with high nucleocytoplasmic ratio).

Blood film image shows the most diagnostic features.

PROVISIONAL DIAGNOSIS

Based on clinical presentations, FBE results, and morphology, this case is highly suggestive of acute myelomonocytic leukaemia and further tests are suggested.

DIFFERENTIAL DIAGNOSES

AML (acute myeloid leukaemia)
ALL (acute lymphoblastic leukaemia)
Acute monoblastic leukaemia
CML (chronic myelocytic leukaemia)
CMML (chronic myelomonocytic leukaemia)
Leukemoid reaction
MDS transformation (myelodysplastic syndrome)
Aplastic anaemia
Infectious mononucleosis and other viral infection

FURTHER TESTS AND EXPECTED RESULTS

Confirmation of Provisional Diagnosis

Bone marrow → more than 20% myeloblasts and more than 20% monoblasts.

Cytochemistry

- **Myeloperoxidase (MPO)** → positive
- **Sudan black B (SBB)** → positive
- **Chloroacetate esterase (CAE)** → positive
- **Alpha naphthyl acetate esterase (ANAE)** → positive

  Immunophenotyping → CD13 & 33 (myeloid markers) and CD11b, 11c, 14, & 64 (monocytic markers) → positive
  Cytogenetic → inv (16) & 11q23

Exclusion of Differential Diagnoses

Cytochemistry

- **Periodic Acid Schiff (PAS)** → negative
- **Acid phosphatase** → negative

  Immunophenotyping → B (CD10, 19, 22) & T (CD 2, 3, 5, 7) cell markers are negative.
  Cytogenetic and molecular studies → t(9:22) ph chr & BCR-ABL → negative to exclude CML

FINAL DIAGNOSIS

Acute myelomonocytic leukaemia (AMML) is the final diagnosis for this case based on further tests that confirm the provisional diagnosis and exclude all the differential diagnoses.

Case 32

Clinical presentations

- Coagulopathy
- Pallor
- Easy bruising
- Neurological symptoms and renal dysfunction

FBE RESULTS

CBC (Complete Blood Count) Parameters

Parameters	Results	Units
Hb	84	g/L
RCC	4.80	$\times 10^{12} \, L^{-1}$
PCV	0.44	
MCV	92	Fl
MCH	18.0	Pg
MCHC	191	g/L
RDW	16.1	%
Retics	N/A	%
WBC	148.0	$\times 10^9 \, L^{-1}$
Plt count	10	$\times 10^9 \, L^{-1}$

Differential White Cell Count

Cell Type	Percentages (%)	Absolute Count
Neutrophils	3	$4.44 \times 10^9 \, L^{-1}$
Bands	0	$0 \times 10^9 \, L^{-1}$
Lymphocytes	1	$1.48 \times 10^9 \, L^{-1}$
Monocytes	6	$8.88 \times 10^9 \, L^{-1}$
Eosinophils	0	$0 \times 10^9 \, L^{-1}$
Basophils	0	$0 \times 10^9 \, L^{-1}$
Metamyelocytes	0	

Continued

Differential White Cell Count—cont'd		
Cell Type	Percentages (%)	Absolute Count
Myelocytes	0	
Promyelocytes	6	
Blasts	84	
Prolymphocytes	0	
nRBC/100WBC	0	

RBC MORPHOLOGY

Mild anisocytosis and irregular contracted cells.

WBC MORPHOLOGY

Large blasts with abundant cytoplasm, some showing vacuolation, nuclear lobulation, and multiple nucleoli.

PLATELET MORPHOLOGY

Marked thrombocytopenia with normal morphology.

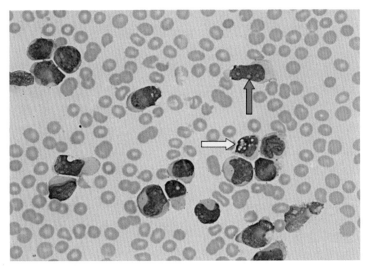

Monoblasts (large with abundant cytoplasm, high N/C ratio, some (showing vacuolation ⬜) nuclear lobulation, and (multiple nucleoli ⬛).

Blood film image shows the most diagnostic features.

PROVISIONAL DIAGNOSIS

Based on clinical presentations, FBE results, and morphology, this case is highly suggestive of acute monoblastic leukaemia and further tests are suggested.

DIFFERENTIAL DIAGNOSES

Other AML (acute myeloid leukaemia)
ALL (acute lymphoblastic leukaemia)
CML (chronic myelocytic leukaemia)
CMML (chronic myelomonocytic leukaemia)
Leukemoid reaction
MDS transformation (myelodysplastic syndrome)
Aplastic anaemia
Infectious mononucleosis and other viral infection

FURTHER TESTS AND EXPECTED RESULTS

Confirmation of Provisional Diagnosis

Bone marrow → more than 80% of non erythroid cells are monoblasts.
Cytochemistry: Alpha naphthyl acetate esterase (ANAE) → positive
Immunophenotyping → CD11b, 11c, 14, & 64 (monocytic markers) → positive
Cytogenetic → 11q23
Molecular studies → MLL/AF-9

Exclusion of Differential Diagnoses

Cytochemistry

- **Periodic Acid Schiff (PAS)** → negative
- **Acid phosphatase** → negative
- **Myeloperoxidase (MPO)** → negative
- **Sudan black B (SBB)** → negative
- **Chloroacetate esterase (CAE)** → negative

Immunophenotyping → B (CD10, 19, 22) & T (CD 2, 3, 5, 7) cell markers are negative
Cytogenetic and molecular studies → t(9:22) **ph chr & BCR-ABL** → negative to exclude CML **(Ph Chr stands for Philadelphia chromosome)**

FINAL DIAGNOSIS

Acute monoblastic leukaemia is the final diagnosis for this case based on further tests that confirm the provisional diagnosis and exclude all the differential diagnoses.

Case 33

Clinical presentations

- Coagulopathy
- Pallor
- Easy bruising
- Neurological symptoms and renal dysfunction

FBE RESULTS

CBC (Complete Blood Count) Parameters

Parameters	Results	Units
Hb	90	g/L
RCC	3.00	$\times 10^{12}\,L^{-1}$
PCV	0.27	
MCV	90	Fl
MCH	33.0	Pg
MCHC	325	g/L
RDW	18.4	%
Retics	N/A	%
WBC	80.0	$\times 10^{9}\,L^{-1}$
Plt count	85	$\times 10^{9}\,L^{-1}$

Differential White Cell Count

Cell Type	Percentages (%)	Absolute Count
Neutrophils	8	$6.4 \times 10^{9}\,L^{-1}$
Bands	9	$7.2 \times 10^{9}\,L^{-1}$
Lymphocytes	6	$4.8 \times 10^{9}\,L^{-1}$
Monocytes	5	$4.0 \times 10^{9}\,L^{-1}$
Eosinophils	2	$1.6 \times 10^{9}\,L^{-1}$
Basophils	0	$0 \times 10^{9}\,L^{-1}$

Continued

Differential White Cell Count—cont'd		
Cell Type	Percentages (%)	Absolute Count
Metamyelocytes	0	
Myelocytes	0	
Promyelocytes	0	
Blasts	40	
Promonocytes	30	
nRBC/100WBC	0	

RBC MORPHOLOGY

Moderate anisocytosis, moderate polychromasia, mild irregular contracted cells, and mild nRBC.

WBC MORPHOLOGY

Moderate monocytosis, mild eosinophilia, and mild left shift.

PLATELET MORPHOLOGY

Marked thrombocytopenia with normal morphology.

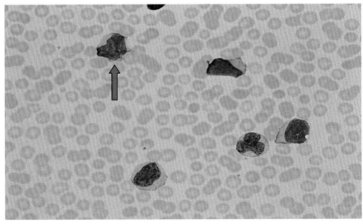

Promonocyte (large, fine reticular pattern, more irregular than blast, up to five nucleoli, and obvious granulated cytoplasm).

Blood film image shows the most diagnostic features.

PROVISIONAL DIAGNOSIS

Based on clinical presentations, FBE results, and morphology, this case is highly suggestive of acute monocytic leukaemia and further tests are suggested.

DIFFERENTIAL DIAGNOSES

Other AML (acute myeloid leukaemia)
ALL (acute lymphoblastic leukaemia)
CML (chronic myelocytic leukaemia)
CMML (chronic myelomonocytic leukaemia)
Leukemoid reaction
MDS transformation (myelodysplastic syndrome)
Aplastic anaemia
Infectious mononucleosis and other viral infection

FURTHER TESTS AND EXPECTED RESULTS

Confirmation of Provisional Diagnosis

Bone marrow → more than 80% of nonerythroid cells are monocytic lineage and the majority of them are promonocyte.
Cytochemistry: Alpha naphthyl acetate esterase (ANAE) → positive
Immunophenotyping → CD11b,11c, 14, & 64 (monocytic markers) → positive
Cytogenetic → 11q23
Molecular studies → MLL/AF-9

Exclusion of Differential Diagnoses

Cytochemistry

- **Periodic Acid Schiff (PAS)** → negative
- **Acid phosphatase** → negative
- **Myeloperoxidase (MPO)** → negative
- **Sudan black B (SBB)** → negative
- **Chloroacetate esterase (CAE)** → negative

Immunophenotyping → B (CD10,19,22) & T (CD 2,3,5,7) cell markers are negative
Cytogenetic and molecular studies → t(9:22) ph chr & BCR-ABL → negative to exclude CML

FINAL DIAGNOSIS

Acute monocytic leukaemia is the final diagnosis for this case based on further tests that confirm the provisional diagnosis and exclude all the differential diagnoses.

Case 34

Clinical presentations

- Fever
- Fatigue
- Easy bruising
- Abdominal pain and hepatosplenomegaly

FBE RESULTS

CBC (Complete Blood Count) Parameters

Parameters	Results	Units
Hb	89	g/L
RCC	2.95	$\times 10^{12}\,L^{-1}$
PCV	0.28	
MCV	94	Fl
MCH	30.0	Pg
MCHC	318	g/L
RDW	17.1	%
Retics	N/A	%
WBC	3.0	$\times 10^{9}\,L^{-1}$
	Corrected to 2.7	
Plt count	30	$\times 10^{9}\,L^{-1}$

Corrected WBC count $= 2.7 \times 10^{9}\,L^{-1}$ because the number of nRBC is $>10\%$.

Differential White Cell Count

Cell Type	Percentages (%)	Absolute Count
Neutrophils	1	$0.03 \times 10^{9}\,L^{-1}$
Bands	10	$0.27 \times 10^{9}\,L^{-1}$
Lymphocytes	79	$2.13 \times 10^{9}\,L^{-1}$
Monocytes	6	$0.16 \times 10^{9}\,L^{-1}$
Eosinophils	0	$0 \times 10^{9}\,L^{-1}$

Continued

Differential White Cell Count—cont'd		
Cell Type	Percentages (%)	Absolute Count
Basophils	0	$0 \times 10^9 \, L^{-1}$
Metamyelocytes	0	
Myelocytes	2	
Promyelocytes	0	
Blasts	2	
Promonocytes	0	
nRBC/100WBC	12	

RBC MORPHOLOGY

Moderate anisocytosis, moderate teardrops, and mild basophilic stippling.

WBC MORPHOLOGY

Mild leucopenia with marked neutropenia.

PLATELET MORPHOLOGY

Marked thrombocytopenia with normal morphology.

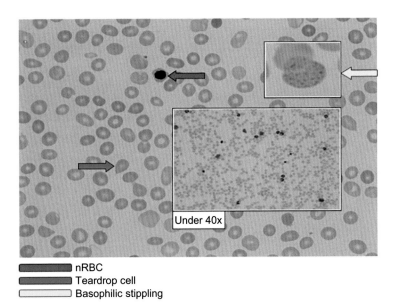

▬▬▬ nRBC
▬▬▬ Teardrop cell
▭▭▭ Basophilic stippling

Blood film image shows the most diagnostic features.

PROVISIONAL DIAGNOSIS

Based on clinical presentations, FBE results, and morphology, this case is highly suggestive of acute erythroid leukaemia and further tests are suggested.

DIFFERENTIAL DIAGNOSES

Other AML (acute myeloid leukaemia)
ALL (acute lymphoblastic leukaemia)
Leukemoid reaction
MDS transformation (myelodysplastic syndrome)
Aplastic anaemia
Pernicious anaemia
Infectious mononucleosis and other viral infection

FURTHER TESTS AND EXPECTED RESULTS

Confirmation of Provisional Diagnosis

Bone marrow → more than 50% of all nucleated cells are erythroblasts and more than 20% of nonerythroid cells are myeloblasts. Erythroid hyperplasia, megaloblastoid changes, dyserythropoietic and dysgranulopoietic changes.

Cytochemistry

- **Periodic Acid Schiff (PAS)** → positive in early erythrocytic precursors
- **Myeloperoxidase (MPO)** → more than 5% positive in myeloblasts
- **Sudan black B (SBB)** → more than 5% positive in myeloblasts

Immunophenotyping →

CD13, 33 & 117 → positive in myeloid component
Glycophorin A → positive in erythroid component

Exclusion of Differential Diagnoses

Cytochemistry: Acid phosphatase → negative
Immunophenotyping → B (CD10,19,22) & T (CD 2,3,5,7) cell markers are negative.
B12 & folate → should be normal to exclude pernicious anaemia

FINAL DIAGNOSIS

Acute erythroid leukaemia is the final diagnosis for this case based on further tests that confirm the provisional diagnosis and exclude all the differential diagnoses.

Case 35

Clinical presentations

- Fever
- Fatigue, weakness, and pallor
- Easy bruising
- Abdominal pain
- Bone tenderness, hepatosplenomegaly, and lymphadenopathy

FBE RESULTS

CBC (Complete Blood Count) Parameters

Parameters	Results	Units
Hb	89	g/L
RCC	3.12	$\times 10^{12} \, L^{-1}$
PCV	0.25	
MCV	80	Fl
MCH	28.5	Pg
MCHC	356	g/L
RDW	20.8	%
Retics	N/A	%
WBC	13.1	$\times 10^9 \, L^{-1}$
Plt count	150	$\times 10^9 \, L^{-1}$

Differential White Cell Count

Cell Type	Percentages (%)	Absolute Count
Neutrophils	47	$6.16 \times 10^9 \, L^{-1}$
Bands	7	$0.92 \times 10^9 \, L^{-1}$
Lymphocytes	10	$1.31 \times 10^9 \, L^{-1}$
Monocytes	1	$0.13 \times 10^9 \, L^{-1}$
Eosinophils	0	$0 \times 10^9 \, L^{-1}$
Basophils	0	$0 \times 10^9 \, L^{-1}$

Differential White Cell Count—cont'd

Cell Type	Percentages (%)	Absolute Count
Metamyelocytes	3	
Myelocytes	2	
Promyelocytes	3	
Blasts	27	
Promonocytes	0	
nRBC/100WBC	0	

RBC MORPHOLOGY

Moderate anisocytosis and mild poikilocytosis cells.

WBC MORPHOLOGY

Large dense blasts with no granules.

PLATELET MORPHOLOGY

Mild giant platelets, thrombocytosis, and megakaryocyte fragments. Platelets budding and clumping.

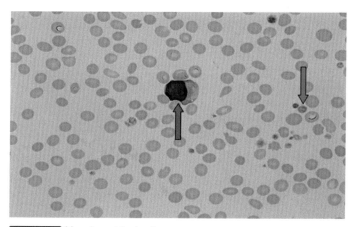

▭ Megakaryoblast cell
Nucleus: round, oval, or kidney shaped, reddish to purple colour, fine, distinct strands to dense chromatin and several nucleoli.
Cytoplasm: moderate to dark blue colour, no granules to fine azurophilic granules in larger blasts.
▭ Giant platelet
Blood film image shows the most diagnostic features.

PROVISIONAL DIAGNOSIS

Based on clinical presentations, FBE results, and morphology, this case is highly suggestive of acute megakaryocytic leukaemia and further tests are suggested.

DIFFERENTIAL DIAGNOSES

Other AML (acute myeloid leukaemia)
ALL (acute lymphoblastic leukaemia)
MDS (myelodysplastic syndrome)
Aplastic anaemia
Chronic myeloid leukaemia (CML)
Bone marrow failure

FURTHER TESTS AND EXPECTED RESULTS

Confirmation of Provisional Diagnosis

Bone marrow →

- Megakaryoblasts are highly pleomorphic
- Increased reticulum fibrosis (often a dry tap)
- More than 20% blasts
- More than 50 megakaryocytic cells

Cytochemistry

- **Periodic Acid Schiff (PAS)** → positive
- **Myeloperoxidase (MPO)** → positive
- **Sudan black B (SBB)** → positive
- **Nonspecific esterase (acetate)** → positive
- **Nonspecific esterase (butyrate)** → negative

Electron microscopy

- **Platelet peroxidase** → positive

Immunophenotyping →

- **CD41 and/or CD61** → positive
- **CD42** → may be positive if the cell is more mature
- **Platelet GPIIIa and von Willebrand factor** → positive
- **CD34 and CD45** → often negative

    Cytogenetic → t(1:22) in some cases

Exclusion of Differential Diagnoses

Cytogenetic and molecular tests →

- **t(9:22) BCR-ABL** → negative exclude CML
- **Genetic abnormalities associated with myelodysplasia such as 5q del** → negative exclude MDS

  Immunophenotype → lymphoid markers such as CD10, 19, 22, 2, 3, and 5 → negative exclude ALL.

FINAL DIAGNOSIS

Acute megakaryocytic leukaemia is the final diagnosis for this case based on further tests that confirm the provisional diagnosis and exclude all the differential diagnoses.

PATHOPHYSIOLOGY OF AML

The underlying pathophysiology in acute myelogenous leukaemia (AML) consists of a maturational arrest of bone marrow cells in the earliest stages of development. The mechanism of this arrest is under study, but in many cases, it involves the activation of abnormal genes through chromosomal translocations and other genetic abnormalities. For example, t(15:17) PML-RARA in acute promyelocytic leukaemia blocks the differentiation and maturation of myeloid cells at promyelocyte stage. This developmental arrest results in two disease processes. First, the production of normal blood cells markedly decreases, which results in varying degrees of anaemia, thrombocytopenia, and neutropenia. Second, the rapid proliferation of these cells, along with a reduction in their ability to undergo programmed cell death (apoptosis), results in their accumulation in the bone marrow, blood, and frequently, the spleen and liver [13].

Case 36

Clinical presentations

- Fever
- Pallor
- Excessive bruising
- Bone pain
- Hepatosplenomegaly and lymphadenopathy

FBE RESULTS

CBC (Complete Blood Count) Parameters

Parameters	Results	Units
Hb	55	g/L
RCC	2.12	$\times 10^{12} \, L^{-1}$
PCV	0.16	
MCV	75	Fl
MCH	26.0	Pg
MCHC	344	g/L
RDW	14.2	%
Retics	N/A	%
WBC	12.8	$\times 10^{9} \, L^{-1}$
Plt count	313	$\times 10^{9} \, L^{-1}$

Differential White Cell Count

Cell Type	Percentages (%)	Absolute Count
Neutrophils	4	$0.51 \times 10^{9} \, L^{-1}$
Bands	1	$0.13 \times 10^{9} \, L^{-1}$
Lymphocytes	39	$4.99 \times 10^{9} \, L^{-1}$
Monocytes	3	$0.38 \times 10^{9} \, L^{-1}$
Eosinophils	1	$0.13 \times 10^{9} \, L^{-1}$

Differential White Cell Count—cont'd

Cell Type	Percentages (%)	Absolute Count
Basophils	0	$0 \times 10^9 \, \text{L}^{-1}$
Metamyelocytes	0	
Myelocytes	0	
Promyelocytes	0	
Blasts	52	
Promonocytes	0	
nRBC/100WBC	0	

RBC MORPHOLOGY

Mildly microcytic and normochromic cells.

WBC MORPHOLOGY

Mild lymphocytosis with increased number of blasts (small, scant cytoplasm, condensed chromatin, no nucleoli, and high N/C ratio).

PLATELET MORPHOLOGY

Normal in number and morphology.

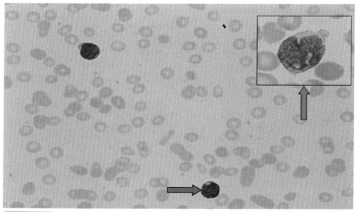

Lymphoblast (small, homogenous, scant cytoplasm, condensed chromatin, no nucleoli, and high N/C ratio): in L1

Lymphoblast (large, heterogeneous, scant cytoplasm, condensed chromatin, two nucleoli, and high N/C ratio): in L2

Blood film image shows the most diagnostic features.

PROVISIONAL DIAGNOSIS

Based on clinical presentations, FBE results, and morphology, this case is highly suggestive of acute lymphoblastic leukaemia and further tests are suggested.

DIFFERENTIAL DIAGNOSES

AML (acute myeloid leukaemia)
Burkitt's lymphoma
Non-Hodgkin lymphoma
Aplastic anaemia
Infectious mononucleosis and other viral infection

FURTHER TESTS AND EXPECTED RESULTS

Confirmation of Provisional Diagnosis

Bone marrow → hyperplasia, more than 25% blasts (small or large (in case of L2), cytoplasm is scant and only slightly or moderately basophilic).

Cytochemistry

- **Periodic Acid Schiff (PAS)** → positive
- **Acid phosphatase** → positive
- **HLA-DR** → positive
- **Terminal deoxynucleotidyl transferase** → positive

  Immunophenotyping → CD34 and TdT (blast) and CD19 & 20 (B-cell markers) → positive in case of precursor-B lymphoblastic leukaemia, or CD34 and TdT (blast) and CD 2, 3, 5, & 7 (T-cell markers) → positive in case of precursor-T lymphoblastic leukaemia.

Cytogenetics →

- t(9;22)(q34;11.2) BCR/ABL → 3%–4%
- t(4;11)(q21;q23) AF4/MLL → 2%–3%
- t(1;19)(q23;p13.3) PBX/E2A → 6%
- t(12;21)(p13;q22) TEL/AML1 → 16%–29%
- Hyperdiploid → 20%–25%
- Hypodiploid → 5%

Exclusion of Differential Diagnoses

Cytochemistry

- **Myeloperoxidase (MPO)** → negative
- **Sudan black B (SBB)** → negative

  Immunophenotyping → CD13 & 33 considered as aberrant markers

FINAL DIAGNOSIS

Acute lymphoblastic leukaemia is the final diagnosis for this case based on further tests that confirm the provisional diagnosis and exclude all the differential diagnoses.

PATHOPHYSIOLOGY OF ALL

In acute lymphoblastic leukaemia, a lymphoid progenitor cell becomes genetically altered and subsequently undergoes dysregulated proliferation, survival, and clonal expansion. In most cases, the pathophysiology of the transformed lymphoid cells reflects the altered expression of genes whose products contribute to the normal development of B-cells and T-cells. Several studies indicate that leukaemic stem cells are present in certain types of acute lymphoblastic leukaemia [14].

Chapter 9

Myeloproliferative/ Myelodysplastic Disorders

Case 37

Clinical presentations

- Anorexia, night sweats, and visual disturbances
- Bone pain
- Anaemia, bleeding disorders, and gout
- Hepatosplenomegaly

FBE RESULTS

CBC (Complete Blood Count) Parameters

Parameters	Results	Units
Hb	127	g/L
RCC	4.16	$\times 10^{12} \, L^{-1}$
PCV	0.39	
MCV	94	Fl
MCH	30.5	Pg
MCHC	325	g/L
RDW	15.3	%
Retics	N/A	%
WBC	53.8	$\times 10^{9} \, L^{-1}$
Plt count	584	$\times 10^{9} \, L^{-1}$

Haematology Case Studies with Blood Cell Morphology and Pathophysiology. http://dx.doi.org/10.1016/B978-0-12-811911-2.00009-X

Differential White Cell Count		
Cell Type	Percentages (%)	Absolute Count
Neutrophils	28	$15.1 \times 10^9 \, L^{-1}$
Bands	16	$8.61 \times 10^9 \, L^{-1}$
Lymphocytes	4	$2.15 \times 10^9 \, L^{-1}$
Monocytes	2	$1.08 \times 10^9 \, L^{-1}$
Eosinophils	4	$2.15 \times 10^9 \, L^{-1}$
Basophils	18	$9.68 \times 10^9 \, L^{-1}$
Metamyelocytes	4	
Myelocytes	15	
Promyelocytes	1	
Blasts	8	
Promonocytes	0	
nRBC/100WBC	0	

RBC MORPHOLOGY

Mild anisocytosis, mild polychromasia, and mild irregular contracted cells.

WBC MORPHOLOGY

Marked leucocytosis, marked left shift, range of immature myeloid cells, 8% blasts, and marked basophilia.

PLATELET MORPHOLOGY

Mild thrombocytosis with normal morphology.

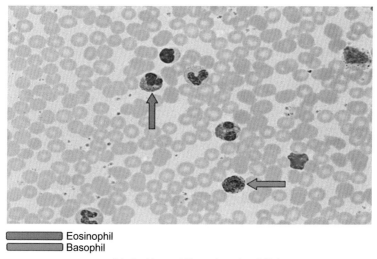

▭ Eosinophil
▭ Basophil

(Marked basophilia and eosinophilia)

Blood film image shows the most diagnostic features.

PROVISIONAL DIAGNOSIS

Based on clinical presentations, FBE results, and morphology, this case is highly suggestive of chronic myeloid leukaemia and further tests are suggested.

DIFFERENTIAL DIAGNOSES

AML (acute myeloid leukaemia)
Reactive neutrophilia
CMML (chronic myelomonocytic leukaemia)
Leukemoid reaction
PV (polycythaemia vera)
ET (essential thrombocythaemia)
MF (myelofibrosis)

FURTHER TESTS AND EXPECTED RESULTS

Confirmation of Provisional Diagnosis

Bone marrow → hypercellular, M:E 10:1–50:1
Cytogenetic → t(9:22) Philadelphia chromosome
Molecular tests → genetic abnormalities (BCR-ABL)
NAP score (neutrophil alkaline phosphatase) → low
LDH, ALP, and uric acid → increased
Vit B12 and transcobalamin I & III → increased

Exclusion of Differential Diagnoses

Bone marrow → blasts less than 20% will exclude AML
JAK2 mutation → negative will exclude (PV, ET, and MF)
NSE → negative will exclude CMML

FINAL DIAGNOSIS

Chronic myeloid leukaemia is the final diagnosis for this case based on further tests that confirm the provisional diagnosis and exclude all the differential diagnoses.

PATHOPHYSIOLOGY OF CML

Chronic myelogenous leukaemia (CML) is an acquired abnormality that involves the hematopoietic stem cell. It is characterized by a cytogenetic aberration consisting of a reciprocal translocation between the long arms of chromosomes 22 and 9; t(9;22). The translocation results in a shortened chromosome 22, an observation first described by Nowell and Hungerford and subsequently termed the Philadelphia (Ph) chromosome after the city of discovery.

This translocation relocates an oncogene called ABL from the long arm of chromosome 9 to the long arm of chromosome 22 in the BCR region. The resulting BCR/ABL fusion gene encodes a chimeric protein with strong tyrosine kinase activity. The expression of this protein leads to the development of the chronic myelogenous leukaemia (CML). The presence of BCR/ABL rearrangement is the hallmark of chronic myelogenous leukaemia (CML), although this rearrangement has also been described in other diseases. It is considered diagnostic when present in a patient with clinical manifestations of CML [15].

Case 38

Clinical presentations

- Headache, confusion, altered mental status, dizziness, visual changes, tinnitus, and paresthesia
- Weight loss, epigastric pain, gout, pruritus, thrombosis, and haemorrhage
- Hepatosplenomegaly and hypertension

FBE RESULTS

CBC (Complete Blood Count) Parameters

Parameters	Results	Units
Hb	208	g/L
RCC	6.60	$\times 10^{12} \, L^{-1}$
PCV	0.65	
MCV	98	Fl
MCH	32.0	Pg
MCHC	320	g/L
RDW	17.4	%
Retics	N/A	%
WBC	10.1	$\times 10^{9} \, L^{-1}$
Plt count	300	$\times 10^{9} \, L^{-1}$

Differential White Cell Count

Cell Type	Percentages (%)	Absolute Count
Neutrophils	79	$7.98 \times 10^{9} \, L^{-1}$
Bands	3	$0.30 \times 10^{9} \, L^{-1}$
Lymphocytes	7	$0.71 \times 10^{9} \, L^{-1}$
Monocytes	11	$1.11 \times 10^{9} \, L^{-1}$
Eosinophils	0	$0 \times 10^{9} \, L^{-1}$
Basophils	0	$0 \times 10^{9} \, L^{-1}$

Continued

Differential White Cell Count—cont'd		
Cell Type	Percentages (%)	Absolute Count
Metamyelocytes	0	
Myelocytes	0	
Promyelocytes	0	
Blasts	0	
Promonocytes	0	
nRBC/100WBC	0	

RBC MORPHOLOGY

Mild anisocytosis, mild macrocytes, and mild polychromasia.

WBC MORPHOLOGY

Normal in number and morphology.

PLATELET MORPHOLOGY

Normal in number and morphology.

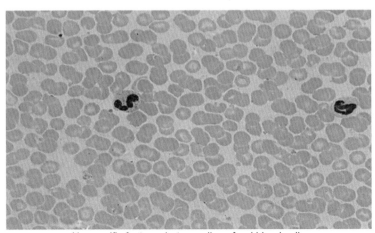

No specific features but crowding of red blood cells

Blood film image shows the most diagnostic features.

PROVISIONAL DIAGNOSES

Based on clinical presentations, FBE results and morphology, this case is highly suggestive of Polycythaemia Vera and further tests are suggested.

DIFFERENTIAL DIAGNOSIS

Methemoglobinaemia
ET (essential thrombocythaemia)
Hypoxia
Renal disease
Iron deficiency anaemia

FURTHER TESTS AND EXPECTED RESULTS

Confirmation of Provisional Diagnosis

Bone marrow →

- Hyperplastic
- Erythroid and granulocytic hyperplasia
- Increased megakaryocytes
- Increased reticulin in postpolycythaemic myelofibrosis and myeloid metaplasia
- Iron stores are often depleted

  Molecular tests → mutation of JAK2 (V617F) & exon 12
  LAP score (leukocyte alkaline phosphatase) → increased

Exclusion of Differential Diagnoses

Oxygen level → normal will exclude hypoxia.
Biochemistry tests → to rule out methemoglobinaemia and renal disease.
Iron studies → normal to rule out iron deficiency anaemia.

FINAL DIAGNOSIS

Polycythaemia vera (PV) is the final diagnosis for this case based on further tests that confirm the provisional diagnosis and exclude all the differential diagnoses.

PATHOPHYSIOLOGY OF PV

Normal stem cells are present in the bone marrow of patients with polycythaemia vera (PV). Also present are abnormal clonal stem cells that interfere with or suppress normal stem cell growth and maturation. Evidence indicates that the aetiology of panmyelosis is unregulated neoplastic proliferation. Progenitors of the blood cells in these patients display abnormal responses to growth factors, suggesting the presence of a defect in a signalling pathway common to different growth factors. The observation that in vitro erythroid colonies grow when no endogenous erythropoietin (Epo) is added to the culture and the presence of a truncated Epo receptor in familial erythrocytosis indicate that the defect is in the transmission of the signal. The sensitivity of polycythaemia vera (PV) progenitors to multiple cytokines suggests that the defect may lie in a common pathway downstream from multiple receptors.

Several reasons suggest that a mutation on the Janus kinase-2 gene (JAK2) is the most likely candidate gene involved in polycythaemia vera (PV) pathogenesis, as JAK2 is directly involved in the intracellular signalling following exposure to cytokines to which polycythaemia vera (PV) progenitor cells display hypersensitivity. A unique valine to phenylalanine substitution at position 617 (V617F) in the pseudokinase JAK2 domain has been identified called JAK2V617F that leads to a permanently turned on signalling at the affected cytokine receptors [16].

Case 39

Clinical presentations

- Epistaxis
- Haemorrhage
- Thrombosis
- Splenomegaly

FBE RESULTS

CBC (Complete Blood Count) Parameters

Parameters	Results	Units
Hb	118	g/L
RCC	4.20	$\times 10^{12}$ L^{-1}
PCV	0.39	
MCV	93	Fl
MCH	28.0	Pg
MCHC	303	g/L
RDW	14.0	%
Retics	N/A	%
WBC	12.0	$\times 10^9$ L^{-1}
Plt count	3400	$\times 10^9$ L^{-1}

Differential White Cell Count

Cell Type	Percentages (%)	Absolute Count
Neutrophils	64	7.68×10^9 L^{-1}
Bands	2	0.24×10^9 L^{-1}
Lymphocytes	27	3.24×10^9 L^{-1}
Monocytes	7	0.48×10^9 L^{-1}
Eosinophils	0	0×10^9 L^{-1}
Basophils	0	0×10^9 L^{-1}

Continued

Differential White Cell Count—cont'd		
Cell Type	Percentages (%)	Absolute Count
Metamyelocytes	0	
Myelocytes	0	
Promyelocytes	0	
Blasts	0	
Promonocytes	0	
nRBC/100WBC	0	

RBC MORPHOLOGY

Moderate anisocytosis, mild polychromasia, moderate irregular contracted cells, and occasional elongated cells.

WBC MORPHOLOGY

Mild neutrophilia and moderate toxic granulation.

PLATELET MORPHOLOGY

Marked thrombocytosis with occasional giant platelets.

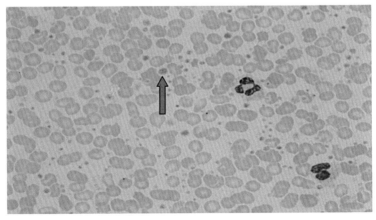

Giant platelet (increased number of platelets)

Blood film image shows the most diagnostic features.

PROVISIONAL DIAGNOSIS

Based on clinical presentations, FBE results, and morphology, this case is highly suggestive of essential thrombocythaemia and further tests are suggested.

DIFFERENTIAL DIAGNOSES

Polycythaemia Vera
Reactive thrombocytosis
Chronic myeloid leukaemia (CML)
Chronic idiopathic myelofibrosis (CIMF)
Myelodysplastic syndrome (MDS)
Iron deficiency anaemia

FURTHER TESTS AND EXPECTED RESULTS

Confirmation of Provisional Diagnosis

Bone marrow →

- Normocellular or slightly hypercellular
- Proliferation of large to giant megakaryocytes and bizarre forms
- Increased megakaryocyte mass
- Increased megakaryocytes arranged in clusters

Molecular tests → mutation of JAK2 (V617F) (50%)

Exclusion of Differential Diagnoses

Bone marrow → no collagen fibrosis will exclude CIMF
Cytogenetics and molecular studies → t(9:22) & BCR-ABL → negative to exclude CML
Cytogenetic abnormality suggestive of MDS such as del 5q → negative to exclude MDS
Iron studies → normal to rule out iron deficiency anaemia
Exclusion the causes of reactive thrombocytosis → normal inflammatory indices

FINAL DIAGNOSIS

Essential thrombocythaemia (ET) is the final diagnosis for this case based on further tests that confirm the provisional diagnosis and exclude all the differential diagnoses.

PATHOPHYSIOLOGY OF ESSENTIAL THROMBOCYTHAEMIA

Essential thrombocythaemia is a myeloproliferative disorder characterized by increased platelet production. In the WHO classification, it is defined as a Ph-negative, BCR-ABL negative condition. Essential thrombocythaemia is predominantly a disease of the middle-aged and elderly population; however, it can also occur in young adults and even in children. The causes of this disorder are not fully known; however, it is thought that genetic mutation involving JAK2 could be the reason behind this disease but it is less frequent compared to PV. Clinical features are either caused directly by the thrombocytosis or reflect the abnormal proliferation of myeloid cells. They include microvascular obstruction, bleeding and, less often, splenomegaly and itch [3].

Case 40

Clinical presentations

- Fatigue, pallor, weight loss, petechiae, and purpura
- Lymphadenopathy
- Hepatosplenomegaly

FBE RESULTS

CBC (Complete Blood Count) Parameters

Parameters	Results	Units
Hb	110	g/L
RCC	3.60	$\times 10^{12} L^{-1}$
PCV	0.35	
MCV	97	Fl
MCH	30.0	Pg
MCHC	314	g/L
RDW	N/A	%
Retics	N/A	%
WBC	11.7	$\times 10^9 L^{-1}$
Plt count	180	$\times 10^9 L^{-1}$

Differential White Cell Count

Cell Type	Percentages (%)	Absolute Count
Neutrophils	48	$5.62 \times 10^9 L^{-1}$
Bands	27	$3.16 \times 10^9 L^{-1}$
Lymphocytes	7	$0.82 \times 10^9 L^{-1}$
Monocytes	3	$0.35 \times 10^9 L^{-1}$
Eosinophils	0	$0 \times 10^9 L^{-1}$
Basophils	0	$0 \times 10^9 L^{-1}$
Metamyelocytes	7	

Continued

Differential White Cell Count—cont'd		
Cell Type	Percentages (%)	Absolute Count
Myelocytes	7	
Promyelocytes	0	
Blasts	1	
Promonocytes	0	
nRBC/100WBC	4	

RBC MORPHOLOGY

Mild polychromasia, moderate teardrops, and occasional nucleated red blood cell.

WBC MORPHOLOGY

Normal in number with mild left shift.

PLATELET MORPHOLOGY

Normal in number and morphology.

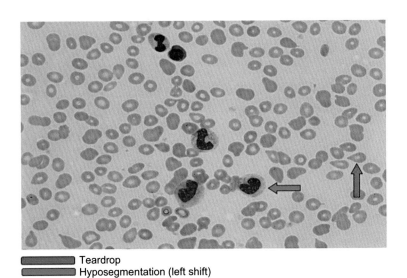

▭ Teardrop
▭ Hyposegmentation (left shift)

Blood film image shows the most diagnostic features.

PROVISIONAL DIAGNOSIS

Based on clinical presentations, FBE results, and morphology, this case is highly suggestive of myelofibrosis and further tests are suggested.

DIFFERENTIAL DIAGNOSES

Acute myeloid leukaemia (AML)
Chronic myeloid leukaemia (CML)
Myelodysplastic syndrome (MDS)
Bone marrow infarction
Tuberculosis (TB)
Acute haemolysis
Severe infection
Systemic lupus erythematosus (SLE)
Acute lymphoblastic leukaemia (ALL)
Chronic lymphocytic leukaemia (CLL)
Lymphoma

FURTHER TESTS AND EXPECTED RESULTS

Confirmation of Provisional Diagnosis

Bone marrow →

- Deceased marrow cellularity
- Increased marrow reticulin and collagen
- Abnormal megakaryocytic proliferation

    Molecular tests → mutation of JAK2 (V617F) (in minority of patients)

Exclusion of Differential Diagnoses

Bone marrow → number of blasts less than 20% exclude AML and ALL with confirmation by cytogenetic, molecular, and immunohistochemistry → exclude other leukaemia
Cytogenetics and molecular studies → t(9:22) and BCR-ABL → negative to exclude CML
Cytogenetic abnormality suggestive of MDS → negative to exclude MDS
Microbiology tests → to exclude TB and other infection
DAT → to exclude the causes of haemolysis
ANA → to exclude SLE

FINAL DIAGNOSIS

Myelofibrosis (MF) is the final diagnosis for this case based on further tests that confirm the provisional diagnosis and exclude all the differential diagnoses.

PATHOPHYSIOLOGY OF MF

The hallmark of myelofibrosis is increased reticulin staining. However, increased reticulin can also be seen in patients with acute leukaemias, especially M7 acute myeloid leukaemia (AML). In a recent report, the extent of myelofibrosis at diagnosis was shown to have prognostic significance in childhood acute lymphoblastic leukaemia (ALL).

The fibrous network observed in myelofibrosis is collagenous and contains fibronectin; the reticulin (silver or Gomori) stain reacts with a protein that is intimately associated with type III collagen and is generally considered to be a form of procollagen. Fibrosis of the bone marrow presumably reflects overgrowth of the normal marrow matrix. Matrix homeostasis results from a balance between its deposition and its removal. The former is regulated by various growth factors, most notably platelet-derived growth factor (PDGF), whereas the latter presumably reflects the activity of collagenase-expressing monocytes, macrophages, and granulocytes. Thus the diseases associated with myelofibrosis can be classified according to whether the basic defect is matrix overproduction, underresorption, or both. The last of these is typified by vitamin D deficiency because $1,25(OH)_2 D_3$, the active metabolite of vitamin D_3, inhibits the proliferation of megakaryocytes and also encourages monocyte/macrophage differentiation. Some investigators believe that the abnormal fibrotic marrow stroma directly enhances the circulation and dissemination of hematopoietic precursors by an unknown mechanism. This leads to extramedullary haematopoiesis in the liver, spleen, lymph nodes, or (occasionally) kidneys, causing myeloid metaplasia in these organs, which then become enlarged. On occasion, hypersplenism may also contribute to cytopenias. Among adults with idiopathic myelofibrosis (IMF), conventional cytogenetic analysis of the marrow reveals an abnormal clone in approximately one-third of patients. Using a comparative genomic hybridization technique, Al-Assar et al. studied IMF marrow specimens and found chromosomal imbalances in 21 of 25 cases. Gains of 9p, 13q, 2q, 3p, and 12q were among the most commonly seen abnormalities. Isolated del(20q) or del(13q) appears to confer a better prognosis. All other abnormalities confer an independent adverse effect on survival and are also associated with higher JAK2V617F mutational frequency. The gain-of-function V617F mutation in the JAK2 gene (on chromosome 9p) is seen in many adult patients with IMF. Its presence correlates with a shift from thrombopoiesis towards increased erythropoiesis and may also predict progression to massive splenomegaly and leukaemic transformation [17].

Case 41

Clinical presentations

- Fatigue and weakness
- Vertigo and loss of appetite
- Haemorrhagic symptoms
- Hepatosplenomegaly
- Skin infiltration

FBE RESULTS

CBC (Complete Blood Count) Parameters

Parameters	Results	Units
Hb	98	g/L
RCC	3.70	$\times 10^{12} \, L^{-1}$
PCV	0.30	
MCV	81	Fl
MCH	26.0	Pg
MCHC	327	g/L
RDW	16.9	%
Retics	N/A	%
WBC	70.0	$\times 10^9 \, L^{-1}$
Plt count	129	$\times 10^9 \, L^{-1}$

Differential White Cell Count

Cell Type	Percentages (%)	Absolute Count
Neutrophils	30	$21.0 \times 10^9 \, L^{-1}$
Bands	1	$0.70 \times 10^9 \, L^{-1}$
Lymphocytes	5	$3.50 \times 10^9 \, L^{-1}$
Monocytes	47	$32.90 \times 10^9 \, L^{-1}$
Eosinophils	0	$0 \times 10^9 \, L^{-1}$
Basophils	0	$0 \times 10^9 \, L^{-1}$

Continued

Differential White Cell Count—cont'd		
Cell Type	Percentages (%)	Absolute Count
Metamyelocytes	1	
Myelocytes	2	
Promyelocytes	14	
Blasts	0	
Prolymphocytes	0	
nRBC/100WBC	0	

RBC MORPHOLOGY

Mild anisocytosis, mild microcytes, and mild hypochromic cells.

WBC MORPHOLOGY

Marked monocytosis, marked neutrophilia with mild hypersegmented neutrophils, and mild hypogranular neutrophils.

PLATELET MORPHOLOGY

Mild thrombocytopenia, mild giant platelets and agranular platelets.

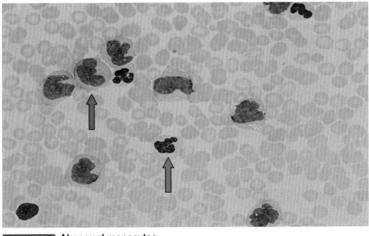

▬ Abnormal monocytes
▬ Hypogranular neutrophil

Blood film image shows the most diagnostic features.

PROVISIONAL DIAGNOSIS

Based on clinical presentations, FBE results, and morphology, this case is highly suggestive of chronic myelomonocytic leukaemia (CMML) and further tests are suggested.

DIFFERENTIAL DIAGNOSES

Chronic myeloid leukaemia (CML) and other MPN
Myelodysplastic syndrome
Tuberculosis (TB)
Bacterial endocarditis

FURTHER TESTS AND EXPECTED RESULTS

Confirmation of Provisional Diagnosis

Bone marrow →

- Hypercellular
- Granulocytic proliferation
- Dyserythropoiesis
- Increased monocytic precursor

Immunophenotype →

- **CD13, 33, & 14** → positive

Cytochemistry →

- **Nonspecific esterase** → positive
- **Myeloperoxidase and Sudan Black B** → negative or weakly positive
- **Periodic Acid-Schiff** → negative

Exclusion of Differential Diagnoses

Cytogenetic and molecular tests →

- **t(9:22) ABL-BCR** → negative exclude CML
- **Cytogenetic abnormality associated with MDS including chromosome 5 & 7** → negative exclude MDS

  Microbiology test for TB and other bacteria → negative exclude TB and bacterial infection.

FINAL DIAGNOSIS

Chronic myelomonocytic leukaemia CMML is the final diagnosis for this case based on further tests that confirm the provisional diagnosis and exclude all the differential diagnoses.

Case 42

Clinical presentations

- Fatigue
- Pallor
- Lethargy
- Breathless

FBE RESULTS

CBC (Complete Blood Count) Parameters

Parameters	Results	Units
Hb	97	g/L
RCC	4.30	$\times 10^{12}\,L^{-1}$
PCV	0.34	
MCV	79	Fl
MCH	23.0	Pg
MCHC	285	g/L
RDW	17.6	%
Retics	N/A	%
WBC	5.7	$\times 10^9\,L^{-1}$
Plt count	120	$\times 10^9\,L^{-1}$

Differential White Cell Count

Cell Type	Percentages (%)	Absolute Count
Neutrophils	56	$3.19 \times 10^9\,L^{-1}$
Bands	0	$0 \times 10^9\,L^{-1}$
Lymphocytes	37	$2.11 \times 10^9\,L^{-1}$
Monocytes	3	$0.17 \times 10^9\,L^{-1}$
Eosinophils	3	$0.17 \times 10^9\,L^{-1}$
Basophils	1	$0.06 \times 10^9\,L^{-1}$

Continued

Differential White Cell Count—cont'd

Cell Type	Percentages (%)	Absolute Count
Metamyelocytes	0	
Myelocytes	0	
Promyelocytes	0	
Blasts	0	
Prolymphocytes	0	
nRBC/100WBC	0	

RBC MORPHOLOGY

Moderate hypochromic microcytes, marked teardrops, mild elongated, and moderate irregular contracted cells.

WBC MORPHOLOGY

Normal in number and morphology.

PLATELET MORPHOLOGY

Mild thrombocytopenia with normal morphology.

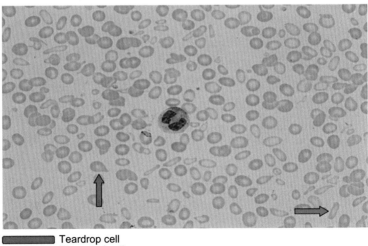

Teardrop cell
Elongated cell

Blood film image shows the most diagnostic features.

PROVISIONAL DIAGNOSIS

Based on clinical presentations, FBE results, and morphology, this case is highly suggestive of refractory anaemia with ringed sideroblasts (RARS) and further tests are suggested.

DIFFERENTIAL DIAGNOSES

Other myelodysplastic syndromes
Chronic myeloid leukaemia (CML)
Hairy cell leukaemia (HCL)
Megaloblastic anaemia
Aplastic anaemia
Bone marrow failure

FURTHER TESTS AND EXPECTED RESULTS

Confirmation of Provisional Diagnosis

Bone marrow →

- Less than 5% blasts
- Dyserythropoiesis
- More than 15% ringed sideroblasts

Exclusion of Differential Diagnoses

Cytochemistry → TRAP → negative exclude hairy cell leukaemia
Cytogenetic and molecular tests → t(9:22)   ABL-BCR → negative exclude CML
B12 & folate → normal → exclude megaloblastic anaemia

FINAL DIAGNOSIS

Refractory anaemia with ringed sideroblasts (RARS) is the final diagnosis for this case based on further tests that confirm the provisional diagnosis and exclude all the differential diagnoses.

Case 43

Clinical presentations

- Fatigue and weakness
- Haemorrhagic symptoms
- Hepatosplenomegaly

FBE RESULTS

CBC (Complete Blood Count) Parameters

Parameters	Results	Units
Hb	91	g/L
RCC	2.60	$\times 10^{12}\,L^{-1}$
PCV	0.29	
MCV	112	Fl
MCH	35.0	Pg
MCHC	314	g/L
RDW	18.2	%
Retics	N/A	%
WBC	6.7	$\times 10^{9}\,L^{-1}$
Plt count	120	$\times 10^{9}\,L^{-1}$

Differential White Cell Count

Cell Type	Percentages (%)	Absolute Count
Neutrophils	56	$3.64 \times 10^{9}\,L^{-1}$
Bands	9	$0.60 \times 10^{9}\,L^{-1}$
Lymphocytes	15	$1.00 \times 10^{9}\,L^{-1}$
Monocytes	10	$0.67 \times 10^{9}\,L^{-1}$
Eosinophils	3	$0.20 \times 10^{9}\,L^{-1}$
Basophils	0	$0 \times 10^{9}\,L^{-1}$
Metamyelocytes	3	

Differential White Cell Count—cont'd

Cell Type	Percentages (%)	Absolute Count
Myelocytes	2	
Promyelocytes	0	
Blasts	2	
Prolymphocytes	0	
nRBC/100WBC	0	

RBC MORPHOLOGY

Moderate macrocytes, mild ovalomacrocytes, mild microcytes, mild polychromasia, and mild teardrops.

WBC MORPHOLOGY

Mild left shift, mild hypogranular neutrophils, and mild pseudo Pelger–Huet.

PLATELET MORPHOLOGY

Mild thrombocytopenia with mild giant platelets.

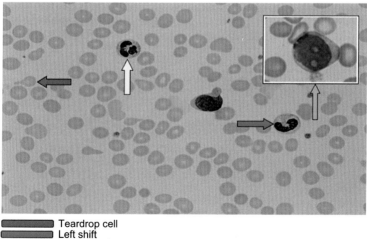

▬ Teardrop cell
▬ Left shift
▭ Hypogranular neutrophil
▭ Blast cell

Blood film image shows the most diagnostic features.

PROVISIONAL DIAGNOSIS

Based on clinical presentations, FBE results, and morphology, this case is highly suggestive of refractory anaemia with excess blasts (RAEB) and further tests are suggested.

DIFFERENTIAL DIAGNOSES

Other myelodysplastic syndromes
Chronic myeloid leukaemia (CML)
Hairy cell leukaemia (HCL)
Megaloblastic anaemia
Aplastic anaemia
Bone marrow failure

FURTHER TESTS AND EXPECTED RESULTS

Confirmation of Provisional Diagnosis

Bone marrow →

- Hypercellular
- Dysgranulopoiesis and/or dyserythropoiesis and/or dysmegakaryopoiesis
- 5%–9% blasts with no Auer rods (RAEB I)
- 10%–19% blasts with Auer rods (RAEB II)

Exclusion of Differential Diagnoses

Cytochemistry → TRAP → negative exclude hairy cell leukaemia
Cytogenetic and molecular tests → t(9:22) ABL-BCR → negative exclude CML
B12 & folate → normal → exclude megaloblastic anaemia

FINAL DIAGNOSIS

Refractory anaemia with excess blasts (RAEB) is the final diagnosis for this case based on further tests that confirm the provisional diagnosis and exclude all the differential diagnoses.

PATHOPHYSIOLOGY OF MDS

Myelodysplastic syndrome (MDS) is a group of heterogeneous diseases which all are the result of abnormal hematopoietic stem cells, initially, at the level of myeloid stem cells. There are different agents that affect the stem cells in MDS. First, a cumulative effect of environmental exposure. Second, chemical insult, viral infection, or radiation that causes mutations. These mutations make the stem cells produce pathologic clones of cells. Abnormality of chromosome 5 and 7 considered to be the most common ones. It is suggested that there is association between MDS and smoking. There are two morphologic findings common in all types of MDS, the presence of progressive cytopenia despite cellular bone marrow and dyspoiesis in one or more cell lines [2].

Chapter 10

Chronic Lymphoproliferative Disorders

Case 44

Clinical presentations

- Enlarged lymph nodes
- Fever, fatigue, weight loss, and night sweats
- Hepatosplenomegaly

FBE RESULTS

CBC (Complete Blood Count) Parameters		
Parameters	Results	Units
Hb	88	g/L
RCC	2.87	$\times 10^{12}\,L^{-1}$
PCV	0.25	
MCV	88	Fl
MCH	30.8	Pg
MCHC	349	g/L
RDW	14.8	%
Retics	N/A	%
WBC	65.2	$\times 10^{9}\,L^{-1}$
Plt count	584	$\times 10^{9}\,L^{-1}$

Haematology Case Studies with Blood Cell Morphology and Pathophysiology. http://dx.doi.org/10.1016/B978-0-12-811911-2.00010-6

Differential White Cell Count

Cell Type	Percentages (%)	Absolute Count
Neutrophils	20	$13.04 \times 10^9 \, L^{-1}$
Bands	0	$0 \times 10^9 \, L^{-1}$
Lymphocytes	80	$52.16 \times 10^9 \, L^{-1}$
Monocytes	0	$0 \times 10^9 \, L^{-1}$
Eosinophils	0	$0 \times 10^9 \, L^{-1}$
Basophils	0	$0 \times 10^9 \, L^{-1}$
Metamyelocytes	0	
Myelocytes	0	
Promyelocytes	0	
Blasts	0	
Promonocytes	0	
nRBC/100WBC	0	

RBC MORPHOLOGY
Mild anisocytosis and mild poikilocytosis.

WBC MORPHOLOGY
Marked leucocytosis, marked lymphocytosis, and marked smear cells.

PLATELET MORPHOLOGY
Marked thrombocytosis with normal morphology.

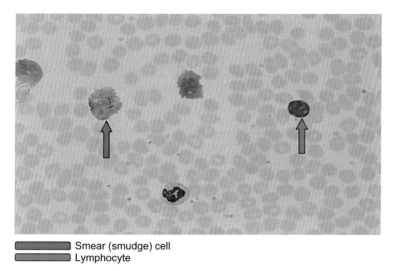

▭ Smear (smudge) cell
▭ Lymphocyte

Blood film image shows the most diagnostic features.

PROVISIONAL DIAGNOSIS

Based on clinical presentations, FBE results, and morphology, this case is highly suggestive of chronic lymphocytic leukaemia and further tests are suggested.

DIFFERENTIAL DIAGNOSES

Acute lymphoblastic leukaemia (ALL)
Acute myeloid leukaemia (AML)
Chronic myeloid leukaemia (CML)
Hairy cell leukaemia (HCL)
Prolymphocytic leukaemia
Lymphomas (diffuse large cell, follicular, lymphoblastic, mantle cell, non-Hodgkin)
Myelodysplastic syndrome (MDS)
Myelofibrosis (MF)
Autoimmune haemolytic anaemia (AIHA)
Postsplenectomy lymphocytosis
Lymphocytosis induced by acute stress
Infectious mononucleosis (IM)
Whooping cough (bacterial infection)

FURTHER TESTS AND EXPECTED RESULTS

Confirmation of Provisional Diagnosis

Bone marrow →

- Replacement by lymphocytes
- Less than 10% of prolymphocytes

Immunophenotype →

- **CD5, 19, 23, & 79** → positive
- **CD20** → weak positive
- **CD10 & FMC7** → negative

Molecular tests →

- del(13q.14.3) ∼ 50%
- trisomy 12 ∼ 20%
- del(11q) ∼ 15%

Exclusion of Differential Diagnoses

Bone marrow → number of blasts less than 20% exclude AML and ALL with confirmation by cytogenetic, molecular, and immunohistochemistry → exclude other leukaemias and lymphomas.
Cytogenetics and molecular studies → t(9:22) & BCR-ABL → negative to exclude CML
Cytogenetic abnormality suggestive of MDS → negative to exclude MDS
Microbiology tests → to exclude bacterial infection
Serology test (monospot) → normal exclude IM
DAT → negative exclude AIHA

FINAL DIAGNOSIS

Chronic lymphocytic leukaemia (CLL) is the final diagnosis for this case based on further tests that confirm the provisional diagnosis and exclude all the differential diagnoses.

PATHOPHYSIOLOGY OF CLL

The cells of origin in the majority of patients with chronic lymphocytic leukaemia (CLL) are clonal B-cells arrested in the B-cell differentiation pathway, intermediate between pre-B-cells and mature B-cells. Morphologically in the peripheral blood, these cells resemble mature lymphocytes. An abnormal karyotype is observed in the majority of patients with chronic lymphocytic leukaemia (CLL). The most common abnormality is deletion of 13q, which occurs in more than 50% of patients.

The presence of trisomy 12, which is observed in 15% of patients, is associated with atypical morphology and progressive disease. Deletion in the short arm of chromosome 17 has been associated with rapid progression, short remission, and decreased overall survival in chronic lymphocytic leukaemia (CLL). 17p13 deletions are associated with loss of function of the tumour suppressor gene p53. Deletions of bands 11q22–q23, observed in 19% of patients, are associated with extensive lymph node involvement, aggressive disease, and shorter survival. Studies have demonstrated that the proto-oncogene BCL2 is overexpressed in B-cell chronic lymphocytic leukaemia (CLL). The proto-oncogene BCL2 is a known suppressor of apoptosis (programmed cell death), therefore the overexpression results in a long life of the involved leukaemic cells [18].

Case 45

Clinical presentations

- Enlarged lymph nodes
- Fever, fatigue, weight loss, and night sweats
- Hepatosplenomegaly

FBE RESULTS

CBC (Complete Blood Count) Parameters

Parameters	Results	Units
Hb	100	g/L
RCC	5.00	$\times 10^{12}\,L^{-1}$
PCV	0.34	
MCV	68	Fl
MCH	20.0	Pg
MCHC	294	g/L
RDW	19.0	%
Retics	14.5	%
WBC	108.0	$\times 10^{9}\,L^{-1}$
Plt count	204	$\times 10^{9}\,L^{-1}$

Differential White Cell Count

Cell Type	Percentages (%)	Absolute Count
Neutrophils	5	$5.4 \times 10^{9}\,L^{-1}$
Bands	4	$4.32 \times 10^{9}\,L^{-1}$
Lymphocytes	89	$96.12 \times 10^{9}\,L^{-1}$
Monocytes	0	$0 \times 10^{9}\,L^{-1}$
Eosinophils	0	$0 \times 10^{9}\,L^{-1}$
Basophils	0	$0 \times 10^{9}\,L^{-1}$
Metamyelocytes	0	

Differential White Cell Count—cont'd

Cell Type	Percentages (%)	Absolute Count
Myelocytes	0	
Promyelocytes	0	
Blasts	0	
Prolymphocytes	2	
nRBC/100WBC	0	

RBC MORPHOLOGY

Moderate anisocytosis, mild macrocytes marked polychromasia, mild sphero-cyte, and mild basophilic stippling.

WBC MORPHOLOGY

Marked leucocytosis, marked lymphocytosis, mild toxic granulation, and marked smear cells.

PLATELET MORPHOLOGY

Normal in number with occasional giant platelets.

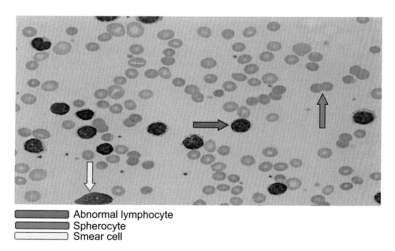

Abnormal lymphocyte
Spherocyte
Smear cell

Blood film image shows the most diagnostic features.

PROVISIONAL DIAGNOSIS

Based on clinical presentations, FBE results, and morphology, this case is highly suggestive of chronic lymphocytic leukaemia combined with autoimmune haemolytic anaemia (CLL + AIHA) and further tests are suggested.

DIFFERENTIAL DIAGNOSES

Acute lymphoblastic leukaemia (ALL)
Acute myeloid leukaemia (AML)
Chronic myeloid leukaemia (CML)
Hairy cell leukaemia (HCL)
Prolymphocytic leukaemia
Lymphomas (diffuse large cell, follicular, lymphoblastic, mantle cell, and non-Hodgkin)
Myelodysplastic syndrome (MDS)
Myelofibrosis (MF)
Postsplenectomy lymphocytosis
Lymphocytosis induced by acute stress
Infectious mononucleosis (IM)
Whooping cough (bacterial infection)

FURTHER TESTS AND EXPECTED RESULTS

Confirmation of Provisional Diagnosis

Bone marrow →

- Replacement by lymphocytes
- Less than 10% of prolymphocytes

Immunophenotype →

- **CD5, 19, 23, & 79** → positive
- **CD20** → weak positive
- **CD10 & FMC7** → negative

Molecular tests →

- del(13q.14.3) ~ 50%
- trisomy 12 ~ 20%
- del(11q) ~ 15%

    DAT → Positive

Exclusion of Differential Diagnoses

Bone marrow → number of blasts less than 20% exclude AML and ALL with confirmation by cytogenetic, molecular, and immunohistochemistry → exclude other leukaemias and lymphomas

Cytogenetics and molecular studies → t(9:22) & BCR-ABL → negative to exclude CML

Cytogenetic abnormality suggestive of MDS → negative to exclude MDS

Microbiology tests → to exclude bacterial infection

Serology test (monospot) → normal exclude IM

FINAL DIAGNOSIS

Chronic lymphocytic leukaemia combined with autoimmune haemolytic anaemia (CLL + AIHA) is the final diagnosis for this case based on further tests that confirm the provisional diagnosis and exclude all the differential diagnoses.

Case 46

Clinical presentations

- Minimal lymphadenopathy
- Fatigue, weight loss, and fever
- Massive hepatosplenomegaly and bone marrow infiltration

FBE RESULTS

CBC (Complete Blood Count) Parameters

Parameters	Results	Units
Hb	128	g/L
RCC	4.50	$\times 10^{12}\,L^{-1}$
PCV	0.41	
MCV	91	Fl
MCH	28.0	Pg
MCHC	312	g/L
RDW	15.1	%
Retics	N/A	%
WBC	85.0	$\times 10^{9}\,L^{-1}$
Plt count	130	$\times 10^{9}\,L^{-1}$

Differential White Cell Count

Cell Type	Percentages (%)	Absolute Count
Neutrophils	5	$4.25 \times 10^{9}\,L^{-1}$
Bands	0	$0 \times 10^{9}\,L^{-1}$
Lymphocytes	10	$8.5 \times 10^{9}\,L^{-1}$
Monocytes	0	$0 \times 10^{9}\,L^{-1}$
Eosinophils	0	$0 \times 10^{9}\,L^{-1}$
Basophils	0	$0 \times 10^{9}\,L^{-1}$
Metamyelocytes	0	

Differential White Cell Count—cont'd		
Cell Type	Percentages (%)	Absolute Count
Myelocytes	0	
Promyelocytes	0	
Blasts	0	
Prolymphocytes	85	
nRBC/100WBC	0	

RBC MORPHOLOGY

Mild anisocytosis.

WBC MORPHOLOGY

Marked lymphocytosis, approximately 85% prolymphocytes with prominent nucleoli and abundant cytoplasm.

PLATELET MORPHOLOGY

Mild thrombocytopenia with normal morphology.

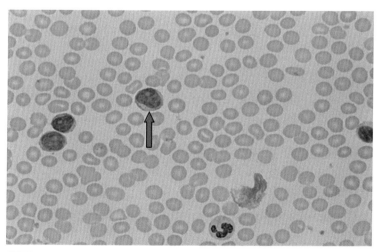

━━━━━ Prolymphocyte (nucleus is round, eccentric, purple, and condensed chromatin whereas cytoplasm is light to blue and has small amount of granules, N/C ratio approximately 3:1, and 1 prominent nucleolus)

Blood film image shows the most diagnostic features.

PROVISIONAL DIAGNOSIS

Based on clinical presentations, FBE results, and morphology, this case is highly suggestive of prolymphocytic leukaemia and further tests are suggested.

DIFFERENTIAL DIAGNOSES

Mantle cell lymphoma in leukaemic phase
Chronic lymphocytic leukaemia in prolymphocytic transformation
Hairy cell leukaemia (HCL)
Splenic lymphoma with villous lymphocyte (SLVL)
Plasma cell leukaemia
Follicular lymphoma
Large cell lymphoma

FURTHER TESTS AND EXPECTED RESULTS

Confirmation of Provisional Diagnosis

Bone marrow →

- Diffusely infiltrated by prolymphocytes
- More than 55% of prolymphocytes

Immunophenotype →

- **CD19, 20, 22, 79a&b, FMC7, & bright sIg** → positive (in case of B-PLL)
- **CD2, 3, 5, 7, & 52** → positive (in case of T-PLL)
- **CD10 & 23** → negative

Cytogenetic tests →

- Trisomy 3
- t(11:14)(q13:q32)
- other rearrangement with 14q32 breakpoint
- TP53 mutation
- Inv(14) (in case of T-PLL)

  Cytochemistry → **acid phosphatase** → positive (in case of T-PLL)

Exclusion of Differential Diagnoses

Immunophenotype → exclude other leukaemia and lymphoma

- **CD138** → negative → exclude plasma cell leukaemia
- **Cyclin D1** → negative → exclude mantle cell lymphoma
- **CD123, 103, & TRAP** → negative → exclude hairy cell leukaemia

FINAL DIAGNOSIS

Prolymphocytic leukaemia (PLL) is the final diagnosis for this case based on further tests that confirm the provisional diagnosis and exclude all the differential diagnoses.

Case 47

Clinical presentations

- Weakness and fatigue
- Easy bruising and bleeding tendency
- Recurrent infections
- Hepatosplenomegaly

FBE RESULTS

CBC (Complete Blood Count) Parameters

Parameters	Results	Units
Hb	97	g/L
RCC	3.20	$\times 10^{12}\,L^{-1}$
PCV	0.33	
MCV	103	Fl
MCH	30.0	Pg
MCHC	294	g/L
RDW	19.1	%
Retics	N/A	%
WBC	26.4	$\times 10^{9}\,L^{-1}$
Plt count	44	$\times 10^{9}\,L^{-1}$

Differential White Cell Count

Cell Type	Percentages (%)	Absolute Count
Neutrophils	4	$1.06 \times 10^{9}\,L^{-1}$
Bands	0	$0 \times 10^{9}\,L^{-1}$
Lymphocytes	10	$2.64 \times 10^{9}\,L^{-1}$
Monocytes	0	$0 \times 10^{9}\,L^{-1}$
Eosinophils	0	$0 \times 10^{9}\,L^{-1}$
Basophils	0	$0 \times 10^{9}\,L^{-1}$
Metamyelocytes	0	

Differential White Cell Count—cont'd

Cell Type	Percentages (%)	Absolute Count
Myelocytes	0	
Promyelocytes	0	
Blasts	0	
Hairy cells	86	
nRBC/100WBC	0	

RBC MORPHOLOGY

Moderate anisocytosis, mild polychromasia, mild macrocytes, mild irregular contracted cells, moderate elongated cells, and occasional nRBC.

WBC MORPHOLOGY

Predominance of lymphoid cells with marked 'hairy' cytoplasmic projections.

PLATELET MORPHOLOGY

Marked thrombocytopenia with normal morphology.

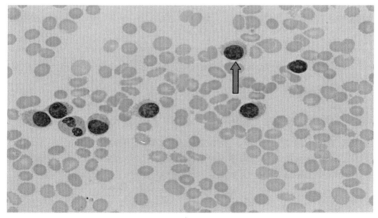

Hairy cell (small-medium sized with oval nucleus, inconspicuous nucleoli, abundant, pale blue cytoplasm with irregular 'hairy' margins).

Blood film image shows the most diagnostic features.

PROVISIONAL DIAGNOSIS

Based on clinical presentations, FBE results, and morphology, this case is highly suggestive of hairy cell leukaemia and further tests are suggested.

DIFFERENTIAL DIAGNOSES

Mantle cell lymphoma (MCL)
Chronic lymphocytic leukaemia (CLL)
Splenic lymphoma with villous lymphocyte (SLVL)
Follicular lymphoma
Prolymphocytic leukaemia (PLL)

FURTHER TESTS AND EXPECTED RESULTS

Confirmation of Provisional Diagnosis

Bone marrow 'dry tap' →

- Many hairy cells and reticulin fibres

Immunophenotype →

- **CD103, 19, 20, 22, 11c, 25, FMC7 & bright sIg** → positive
- **CD21** → negative

Cytogenetic tests →

- Abnormality of chromosome 5
- Monosomy 10
- Monosomy 17

Cytochemistry →

- **Tartrate-resistant acid phosphatase (TRAP)** → positive
- **Specific esterase (naphthol AS-D chloroacetate esterase) and myeloperoxidase** → negative

Exclusion of Differential Diagnoses

Immunophenotype → exclude other leukaemia and lymphoma

- **CD10** → negative → might exclude follicular lymphoma
- **Cyclin D1** → negative → exclude mantle cell lymphoma

FINAL DIAGNOSIS

Hairy cell leukaemia (HCL) is the final diagnosis for this case based on further tests that confirm the provisional diagnosis and exclude all the differential diagnoses.

PATHOPHYSIOLOGY OF HCL

The abnormal cell in hairy cell leukaemia is a clonal B-cell lymphocyte. This cell infiltrates the patient's reticuloendothelial system and interferes with bone marrow function, resulting in bone marrow failure or pancytopenia. The hairy cell also infiltrates the liver and spleen, resulting in organomegaly. The aetiology of hairy cell leukaemia has not been determined, although some investigators suggest that exposures to benzene, organophosphorus insecticides, or other solvents may be related to disease development. Exposure to radiation, agricultural chemicals, and wood dust, and a previous history of infectious mononucleosis have been suggested as aetiologic associations in previous reports. Overexpression of cyclin D1 protein, an important cell cycle regulator, has been observed in hairy cell leukaemia and may play a role in the molecular pathogenesis of the disease. Accumulation of hairy cells in the bone marrow, liver, and spleen, with very little lymph node involvement, is characteristic of hairy cell leukaemia. This pattern probably results from the expression of the integrin receptor alpha4-beta1 by the hairy cells and the interaction of the receptor with the vascular adhesion molecule-1 (VCAM-1) found in splenic and hepatic endothelia, bone marrow, and splenic stroma [19].

Chapter 11

Lymphomas

Case 48

Clinical presentations

- Fever of unknown origin
- Fatigue
- Night sweats and weight loss
- Lymphadenopathy
- Purpura

FBE RESULTS

CBC (Complete Blood Count) Parameters

Parameters	Results	Units
Hb	100	g/L
RCC	4.30	$\times 10^{12}\,L^{-1}$
PCV	0.42	
MCV	89	Fl
MCH	23.0	Pg
MCHC	238	g/L
RDW	N/A	%
Retics	N/A	%
WBC	15.2	$\times 10^{9}\,L^{-1}$
Plt count	5	$\times 10^{9}\,L^{-1}$

Haematology Case Studies with Blood Cell Morphology and Pathophysiology. http://dx.doi.org/10.1016/B978-0-12-811911-2.00011-8

Differential White Cell Count

Cell Type	Percentages (%)	Absolute Count
Neutrophils	7	$1.06 \times 10^9 \, L^{-1}$
Bands	13	$1.98 \times 10^9 \, L^{-1}$
Lymphocytes	14	$2.13 \times 10^9 \, L^{-1}$
Monocytes	6	$0.91 \times 10^9 \, L^{-1}$
Eosinophils	0	$0 \times 10^9 \, L^{-1}$
Basophils	0	$0 \times 10^9 \, L^{-1}$
Metamyelocytes	2	
Myelocytes	5	
Promyelocytes	0	
Blasts	53	
Promonocytes	0	
nRBC/100WBC	0	

RBC MORPHOLOGY

Mild anisocytosis with normocytic normochromic cells.

WBC MORPHOLOGY

Basophilic blasts—very pleomorphic with some showing fine vacuolation. Large irregular nucleus with prominent, singular nucleoli.

PLATELET MORPHOLOGY

Marked thrombocytopenia with normal morphology.

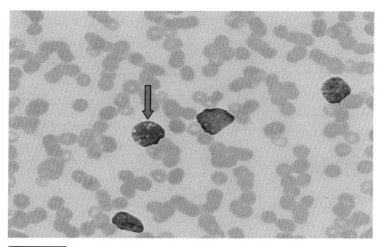

▬▬▬▬ Large blast with very basophilic vacuolated cytoplasm

Blood film image shows the most diagnostic features.

PROVISIONAL DIAGNOSIS

Based on clinical presentations, FBE results, and morphology, this case is highly suggestive of Burkett's lymphoma and further tests are suggested.

DIFFERENTIAL DIAGNOSES

> ALL (L1&2)
> Mantle cell lymphoma (MCL)
> Diffuse large B-cell lymphoma
> Prolymphocytic leukaemia (PLL)

FURTHER TESTS AND EXPECTED RESULTS

Confirmation of Provisional Diagnosis

> **Bone marrow** → hyperplasia, more than 25% blasts (large, very basophilic vacuolated cytoplasm).

Cytochemistry

- **Oil red O** → positive
- **HLA-DR** → positive
- **Terminal deoxynucleotidyl transferase** → negative

> **Immunophenotyping** → CD10, 19, 20, 22, FMC7, BCL6, 79a & sIg → positive

Cytogenetics →

- t(8;14)(q24;q32) c-MYC & Ig heavy chain locus → 75%
- t(2;8)(q13;q24), t(8;22)(q24;q11) c-MYC & kappa or lambda light chain locus → 25%

Exclusion of Differential Diagnoses

Cytochemistry

- **Periodic Acid Schiff (PAS) & TdT** → negative → exclude ALL1&2

Immunophenotyping →

- **BCL1** → negative → exclude mantle cell lymphoma
- **CD23** → negative → exclude PLL
- **CD5** → negative → exclude PLL and MCL

FINAL DIAGNOSIS

Burkett's lymphoma is the final diagnosis for this case based on further tests that confirm the provisional diagnosis and exclude all the differential diagnoses.

Case 49

Clinical presentations

- Tiredness
- Loss of appetite and loss weight
- Enlarged lymph nodes
- Fever and night sweats

FBE RESULTS

CBC (Complete Blood Count) Parameters

Parameters	Results	Units
Hb	95	g/L
RCC	3.08	$\times 10^{12} \, L^{-1}$
PCV	0.28	
MCV	90	Fl
MCH	31.0	Pg
MCHC	343	g/L
RDW	16.9	%
Retics	N/A	%
WBC	3.5	$\times 10^{9} \, L^{-1}$
Plt count	46	$\times 10^{9} \, L^{-1}$

Differential White Cell Count

Cell Type	Percentages (%)	Absolute Count
Neutrophils	37	$1.29 \times 10^{9} \, L^{-1}$
Bands	4	$0.14 \times 10^{9} \, L^{-1}$
Lymphocytes	52	$1.82 \times 10^{9} \, L^{-1}$
Monocytes	5	$0.17 \times 10^{9} \, L^{-1}$
Eosinophils	0	$0 \times 10^{9} \, L^{-1}$
Basophils	1	$0.03 \times 10^{9} \, L^{-1}$

Continued

Differential White Cell Count—cont'd		
Cell Type	Percentages (%)	Absolute Count
Metamyelocytes	0	
Myelocytes	0	
Promyelocytes	0	
Blasts	1	
Prolymphocytes	0	
nRBC/100WBC	0	

RBC MORPHOLOGY

Normocytic normochromic cells and mild polychromasia.

WBC MORPHOLOGY

Moderate toxic granulation and mild lymphoma cells.

PLATELET MORPHOLOGY

Marked thrombocytopenia with normal morphology.

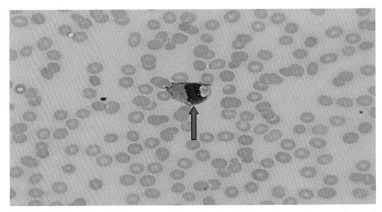

Mantle cell (large to medium-sized cell, irregular nucleus, and scant cytoplasm).

Blood film image shows the most diagnostic features.

PROVISIONAL DIAGNOSIS

Based on clinical presentations, FBE results, and morphology, this case is highly suggestive of chronic mantle cell lymphoma and further tests are suggested.

DIFFERENTIAL DIAGNOSES

    Chronic lymphocytic leukaemia (CLL)
    Prolymphocytic leukaemia (PLL)
    Hairy cell leukaemia (HCL)
    Diffuse large cell lymphoma
    Follicular lymphoma

FURTHER TESTS AND EXPECTED RESULTS

Confirmation of Provisional Diagnosis

Immunophenotype →

- **CD43, cyclin D1, sIg, CD5 & BCL2** → positive

Cytogenetic →

- t(11:14)(q13:32)

Exclusion of Differential Diagnoses

Immunophenotype →

- **CD10** → negative exclude follicular lymphoma
- **CD103** → negative exclude HCL
- **CD23** → negative exclude PLL

FINAL DIAGNOSIS

Chronic mantle cell lymphoma (MCL) is the final diagnosis for this case based on further tests that confirm the provisional diagnosis and exclude all the differential diagnoses.

PATHOPHYSIOLOGY OF MANTLE CELL LYMPHOMA

MCL is a lymphoproliferative disorder derived from a subset of naive pregerminal centre cells localized in primary follicles or in the mantle region of secondary follicles. Most cases of MCL are associated with chromosome translocation t(11;14)(q13;q32). This translocation involves the immunoglobulin heavy chain gene on chromosome 14 and the BCL1 locus on chromosome 11. The molecular consequence of this translocation is the overexpression of protein cyclin D1. Cyclin D1 plays a key role in cell cycle regulation and progression of cells from G1 phase to S phase by activation of cyclin-dependent kinases [20].

Chapter 12

Plasma Cell Disorders

Case 50

Clinical presentations

- Hepatosplenomegaly
- Lymphadenopathy
- Lytic bone lesions
- Severe anaemia

FBE RESULTS

CBC (Complete Blood Count) Parameters

Parameters	Results	Units
Hb	101	g/L
RCC	3.37	$\times 10^{12}\,L^{-1}$
PCV	0.30	
MCV	88	Fl
MCH	30.0	Pg
MCHC	330	g/L
RDW	N/A	%
Retics	N/A	%
WBC	7.0	$\times 10^{9}\,L^{-1}$
Plt count	150	$\times 10^{9}\,L^{-1}$

Differential White Cell Count

Cell Type	Percentages (%)	Absolute Count
Neutrophils	17	$1.19 \times 10^9 \, L^{-1}$
Bands	2	$0.14 \times 10^9 \, L^{-1}$
Lymphocytes	10	$0.70 \times 10^9 \, L^{-1}$
Monocytes	0	$0 \times 10^9 \, L^{-1}$
Eosinophils	1	$0.07 \times 10^9 \, L^{-1}$
Basophils	0	$0 \times 10^9 \, L^{-1}$
Metamyelocytes	0	
Myelocytes	0	
Promyelocytes	0	
Blasts	0	
Plasma cells	70	
nRBC/100WBC	0	

RBC MORPHOLOGY

Moderate anisocytosis, mild polychromasia, mild elongated cells, occasional irregular contracted cells, occasional nucleated red blood cells, and moderate rouleaux.

WBC MORPHOLOGY

Moderate neutropenia, mild toxic granulation, and marked plasma cells.

PLATELET MORPHOLOGY

Normal in number and morphology.

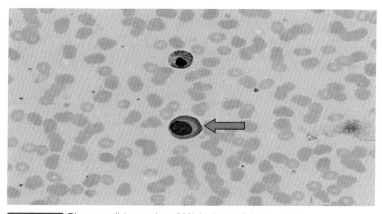

Plasma cell (more than 20% in the peripheral blood)

Blood film image shows the most diagnostic features.

PROVISIONAL DIAGNOSIS

Based on clinical presentations, FBE results, and morphology, this case is highly suggestive of plasma cell leukaemia and further tests are suggested.

DIFFERENTIAL DIAGNOSES

Chronic lymphocytic leukaemia (CLL)
Plasma cell lymphoma (multiple myeloma)
Waldenstrom's macroglobinaemia

FURTHER TESTS AND EXPECTED RESULTS

Confirmation of Provisional Diagnosis

Bone marrow →

- Diffuse plasma cell infiltration (50%–100% of cells are plasma cells)
- Plasma cells are well differentiated
- Binucleated plasma cells may be present

Immunophenotype →

- **CD38, 20, 79a & CD138** → positive

Cytogenetic →

- Monosomy 13
- Gains or losses in chromosome 1

- Trisomy18
- Monosomy X in women
- t(11:14)(q13:32)

Chemistries →

- Increased serum calcium
- Increased blood urea nitrogen level
- Increased creatinine level
- Protein electrophoresis (demonstration of monoclonal paraprotein)

Exclusion of Differential Diagnoses

Immunophenotype →

- **CD58 & CD56** → negative exclude plasma cell myeloma
- **IgM** → normal may exclude Waldenstrom's macroglobinaemia

FINAL DIAGNOSIS

Plasma cell leukaemia (PCL) is the final diagnosis for this case based on further tests that confirm the provisional diagnosis and exclude all the differential diagnoses.

Case 51

Clinical presentations

- Fatigue
- Bone pain and lytic bone lesions
- Neurologic abnormalities
- Infections and renal failure

FBE RESULTS

CBC (Complete Blood Count) Parameters

Parameters	Results	Units
Hb	90	g/L
RCC	2.72	$\times 10^{12} \, L^{-1}$
PCV	0.27	
MCV	98	Fl
MCH	33.0	Pg
MCHC	333	g/L
RDW	16.8	%
Retics	N/A	%
WBC	3.0	$\times 10^9 \, L^{-1}$
Plt count	20	$\times 10^9 \, L^{-1}$

Differential White Cell Count

Cell Type	Percentages (%)	Absolute Count
Neutrophils	67	$2.01 \times 10^9 \, L^{-1}$
Bands	11	$0.33 \times 10^9 \, L^{-1}$
Lymphocytes	20	$0.60 \times 10^9 \, L^{-1}$
Monocytes	1	$0.03 \times 10^9 \, L^{-1}$
Eosinophils	0	$0 \times 10^9 \, L^{-1}$
Basophils	0	$0 \times 10^9 \, L^{-1}$

Continued

Differential White Cell Count—cont'd		
Cell Type	Percentages (%)	Absolute Count
Metamyelocytes	0	
Myelocytes	0	
Promyelocytes	0	
Blasts	0	
Plasma cells	1	
nRBC/100WBC	0	

RBC MORPHOLOGY

Moderate anisocytosis, moderate macrocytes, mild polychromasia, occasional nucleated red blood cells, and marked rouleaux.

WBC MORPHOLOGY

Mild leucopenia with mild toxic granulation.

PLATELET MORPHOLOGY

Marked thrombocytopenia with normal morphology.

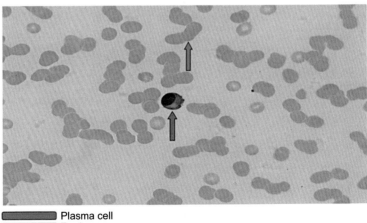

Plasma cell
Rouleaux

Blood film image shows the most diagnostic features.

PROVISIONAL DIAGNOSIS

Based on clinical presentations, FBE results, and morphology, this case is highly suggestive of plasma cell myeloma and further tests are suggested. The DNA content and cytogenetic characteristics of plasma cell leukaemia and plasma cell myeloma are different leading to a different disease evolution. The DNA content in plasma cell leukaemia is normal whereas it is hyperdiploid in plasma cell myeloma.

DIFFERENTIAL DIAGNOSES

Reactive plasmacytosis
Plasma cell leukaemia
Waldenstrom's macroglobinaemia
Monoclonal gammopathies of uncertain origin

FURTHER TESTS AND EXPECTED RESULTS

Confirmation of Provisional Diagnosis

Bone marrow →

- Marrow plasmacytosis (more than 10%)
- Myeloma cells are present (single eccentrically placed nucleus, nucleoli may be seen in finely divided chromatin and various type of inclusions may be present).

Immunophenotype →

- **CD38, 56, 58 & CD138** → positive

Cytogenetic/FISH studies →

- Deletion/loss 13q
- 17p/p53 deletion
- t(4:14)
- t(14:16)

Chemistries →

- Increased total protein
- IgG, IgA, IgE, or IgD are increased
- Presence of monoclonal paraprotein or M protein band by using electrophoresis

- Presence of Bence Jones protein in the urine
- Increased serum calcium
- Increased blood urea nitrogen level
- Increased creatinine level
- Decreased serum albumin with advanced disease

 X rays, CT scan, or MRI → to demonstrate the lytic lesions in the skull and other bones

Exclusion of Differential Diagnoses

Immunophenotype →

- **Light chain restriction** → positive exclude reactive plasmacytosis

FINAL DIAGNOSIS

Plasma cell myeloma is the final diagnosis for this case based on further tests that confirm the provisional diagnosis and exclude all the differential diagnoses.

PATHOPHYSIOLOGY OF PLASMA CELL MYELOMA

Plasma cell myeloma can cause a wide variety of problems. The proliferation of plasma cells may interfere with the normal production of blood cells, resulting in leucopenia, anaemia, and thrombocytopenia. The cells may cause soft-tissue masses (plasmacytomas) or lytic lesions in the skeleton. Feared complications of plasma cell myeloma are bone pain, hypercalcaemia, renal failure, and spinal cord compression. The aberrant antibodies that are produced lead to impaired humoral immunity, and patients have a high prevalence of infection, especially with encapsulated organisms such as *Pneumococcus*. The overproduction of these antibodies may lead to hyperviscosity, amyloidosis, and renal failure [21].

Chapter 13

Haemostatic Disorders (Microangiopathic Haemolytic Anaemia)

Case 52

Clinical presentations
- Mucus membrane bleeding
- Extensive bruising
- Renal failure
- Gangrene

FBE RESULTS

CBC (Complete Blood Count) Parameters		
Parameters	Results	Units
Hb	94	g/L
RCC	3.40	$\times 10^{12} \, L^{-1}$
PCV	0.29	
MCV	85	Fl
MCH	28.0	Pg
MCHC	324	g/L
RDW	18.1	%
Retics	N/A	%
WBC	17.3	$\times 10^{9} \, L^{-1}$
Plt count	30	$\times 10^{9} \, L^{-1}$

Differential White Cell Count		
Cell Type	Percentages (%)	Absolute Count
Neutrophils	63	$10.90 \times 10^9\ L^{-1}$
Bands	13	$2.25 \times 10^9\ L^{-1}$
Lymphocytes	5	$0.87 \times 10^9\ L^{-1}$
Monocytes	8	$1.38 \times 10^9\ L^{-1}$
Eosinophils	1	$0.17 \times 10^9\ L^{-1}$
Basophils	1	$0.17 \times 10^9\ L^{-1}$
Metamyelocytes	2	
Myelocytes	7	
Promyelocytes	0	
Blasts	0	
Prolymphocytes	0	
nRBC/100WBC	2	

RBC MORPHOLOGY

Moderate anisocytosis, mild polychromasia, moderate irregular contracted cells, and moderate fragments.

WBC MORPHOLOGY

Moderate leucocytosis, moderate neutrophilia, moderate left shift, and marked toxic granulation.

PLATELET MORPHOLOGY

Marked thrombocytopenia with occasional giant platelets.

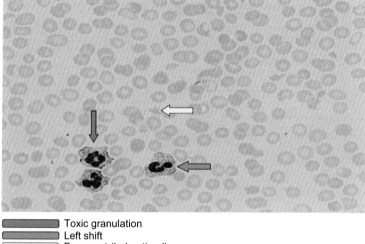

- Toxic granulation
- Left shift
- Fragment (helmet) cell

Blood film image shows the most diagnostic features.

PROVISIONAL DIAGNOSIS

Based on clinical presentations, FBE results, and morphology, this case is highly suggestive of disseminated intravascular coagulation and further tests are suggested.

DIFFERENTIAL DIAGNOSES

- Idiopathic thrombocytopenic purpura
- Haemolytic uremic syndrome
- Thrombotic thrombocytopenic purpura
- Severe liver disease
- Malignancy
- Drug-induced thrombocytopenia
- Sepsis

FURTHER TESTS AND EXPECTED RESULTS

Confirmation of Provisional Diagnosis

Coagulation tests

- **PT** → prolonged
- **aPTT** → prolonged
- **TT** → increased
- **Fibrinogen** → decreased
- **D-dimer** → increased

Exclusion of Differential Diagnoses

ADAMTS13 → normal
Drug history → negative
Renal function tests → normal
Blood culture *E. coli* → negative
Microbiology test for bacteria → negative

FINAL DIAGNOSIS

Disseminated intravascular coagulation (DIC) is the final diagnosis for this case based on further tests that confirm the provisional diagnosis and exclude all the differential diagnoses.

PATHOPHYSIOLOGY OF DIC

DIC is caused by widespread and ongoing activation of coagulation, leading to vascular or microvascular fibrin deposition, thereby compromising an adequate blood supply to various organs. Four different mechanisms are primarily responsible for the haematologic derangements seen in DIC: increased thrombin generation, a suppression of anticoagulant pathways, impaired fibrinolysis, and inflammatory activation. Activation of intravascular coagulation is mediated almost entirely by the intrinsic clotting pathway. Exposure to tissue factor in the circulation occurs via endothelial disruption, tissue damage, or inflammatory or tumour cell expression of procoagulant molecules, including tissue factor. Tissue factor activates coagulation by the extrinsic pathway involving factor VIIa. Factor VIIa has been implicated as the central mediator of intravascular coagulation in sepsis. Blocking the factor VIIa pathway in sepsis has been shown to prevent the development of DIC, whereas interrupting alternative pathways did not demonstrate any effect on clotting. The tissue factor–VIIa complex then serves to activate thrombin, which, in turn, cleaves fibrinogen to fibrin while simultaneously causing platelet aggregation. Evidence suggests that the intrinsic (or contact) pathway is also activated in DIC, while contributing more to hemodynamic instability and hypotension than to activation of clotting. Thrombin generation is usually tightly regulated by multiple haemostatic mechanisms. However, once intravascular coagulation commences, compensatory mechanisms are overwhelmed or incapacitated. Antithrombin is one such mechanism responsible for regulating thrombin levels. However, due to multiple factors, antithrombin activity is reduced in patients with sepsis. First, antithrombin is continuously consumed by ongoing activation of coagulation. Moreover, elastase produced by activated neutrophils degrades antithrombin as well as other proteins. Further antithrombin is lost to capillary leakage. Lastly, production of antithrombin is impaired secondary to liver damage resulting from under-perfusion and microvascular coagulation. Decreased levels of antithrombin have been shown to correlate with elevated mortality in patients with sepsis.

Protein C, along with protein S, serves as an important anticoagulant compensatory mechanism. Under normal conditions, protein C is activated by thrombin and is complexed on the endothelial cell surface with thrombomodulin. Activated protein C combats coagulation via proteolytic cleavage of factors Va and VIIIa. However, the cytokines (tumour necrosis factor α [TNF-a], interleukin 1 [IL-1]) produced in sepsis and other generalized inflammatory states largely incapacitate the protein C pathway. Inflammatory cytokines downregulate the expression of thrombomodulin on the endothelial cell surface. Protein C levels are further reduced via consumption, extravascular leakage, and reduced hepatic production and by a reduction in freely circulating protein S. Tissue factor pathway inhibitor (TFPI) is another anticoagulant mechanism that is disabled in DIC. TFPI inhibits the tissue factor–VIIa complex. Although levels of TFPI are normal in patients with sepsis, a relative insufficiency in DIC is evident. TFPI depletion in animal models predisposes to DIC, and TFPI blocks the procoagulant effect of endotoxin in humans. The intravascular fibrin produced by thrombin is normally eliminated via a process termed fibrinolysis. The initial response to inflammation appears to be augmentation of fibrinolytic action; however, this response soon reverses as inhibitors (plasminogen activator inhibitor-1 [PAI-1], TAFI) of fibrinolysis are released. Indeed, high levels of PAI-1 precede DIC and predict poor outcomes. Fibrinolysis cannot keep pace with increased fibrin formation, eventually resulting in underopposed fibrin deposition in the vasculature. Inflammatory and coagulation pathways interact in substantial ways. Many of the activated coagulation factors produced in DIC contribute to the propagation of inflammation by stimulating endothelial cell release of proinflammatory cytokines. Factor Xa, thrombin, and the tissue factor–VIIa complex have each been demonstrated to elicit proinflammatory action. Furthermore, given the antiinflammatory action of activated protein C and AT, their impairment in DIC contributes to further dysregulation of inflammation. The components of DIC include exposure of blood to procoagulant substances, fibrin deposition in the microvasculature, impaired fibrinolysis, depletion of coagulation factors and platelets (consumptive coagulopathy), and organ damage and failure [22].

Case 53

Clinical presentations

- Weakness
- Purpura
- Gastroenteritis
- Diarrhoea

FBE RESULTS

CBC (Complete Blood Count) Parameters

Parameters	Results	Units
Hb	64	g/L
RCC	2.90	$\times 10^{12}\,L^{-1}$
PCV	0.25	
MCV	86	Fl
MCH	22.0	Pg
MCHC	256	g/L
RDW	21.3	%
Retics	N/A	%
WBC	21.0	$\times 10^9\,L^{-1}$
Plt count	80	$\times 10^9\,L^{-1}$

Differential White Cell Count

Cell Type	Percentages (%)	Absolute Count
Neutrophils	24	$5.04 \times 10^9\,L^{-1}$
Bands	9	$1.89 \times 10^9\,L^{-1}$
Lymphocytes	53	$11.13 \times 10^9\,L^{-1}$
Monocytes	9	$1.89 \times 10^9\,L^{-1}$
Eosinophils	1	$0.21 \times 10^9\,L^{-1}$
Basophils	1	$0.21 \times 10^9\,L^{-1}$

Differential White Cell Count—cont'd

Cell Type	Percentages (%)	Absolute Count
Metamyelocytes	0	
Myelocytes	3	
Promyelocytes	0	
Blasts	0	
Prolymphocytes	0	
nRBC/100WBC	0	

RBC MORPHOLOGY

Marked anisocytosis, moderate polychromasia, marked fragments, marked crenated cells, mild spherocytes, marked acanthocytes, and occasional nRBC.

WBC MORPHOLOGY

Marked neutrophilia, mild lymphocytosis, mild reactive lymphocytes, moderate monocytosis, and mild left shift.

PLATELET MORPHOLOGY

Moderate thrombocytopenia with normal morphology.

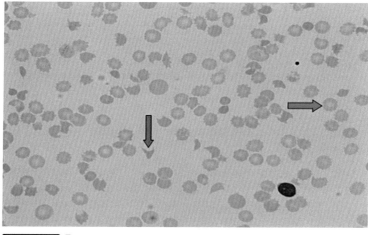

▭ Fragment cell
▭ Crenated cell

Blood film image shows the most diagnostic features.

PROVISIONAL DIAGNOSIS

Based on clinical presentations, FBE results, and morphology, this case is highly suggestive of haemolytic uremic syndrome and further tests are suggested.

DIFFERENTIAL DIAGNOSES

- Idiopathic thrombocytopenic purpura
- Disseminated intravascular coagulation
- Thrombotic thrombocytopenic purpura
- Sepsis/bacterial
- Haemolysis elevated liver enzyme and low platelets (HELLP)
- Drug-induced thrombocytopenia

FURTHER TESTS AND EXPECTED RESULTS

Confirmation of Provisional Diagnosis

Renal function tests → abnormal
Blood culture _E. coli_ → positive

Exclusion of Differential Diagnoses

ADAMTS13 → normal
Coagulation tests (PT, aPTT, D-dimer, and fibrinogen) → normal
Drug history → negative
Microbiology test for sepsis → negative

FINAL DIAGNOSIS

Haemolytic uremic syndrome (HUS) is the final diagnosis for this case based on further tests that confirm the provisional diagnosis and exclude all the differential diagnoses.

PATHOPHYSIOLOGY OF HAEMOLYTIC UREMIC SYNDROME (HUS)

Haemolytic uremic syndrome (HUS) falls into the broader category of thrombotic microangiopathies (TMA). Thrombotic microangiopathies are characterized by the involvement of widespread occlusive microvascular thromboses resulting in thrombocytopenia, microangiopathic haemolytic anaemia, and variable signs and symptoms of end-organ ischaemia.

Two predominant types of HUS are identified: one type involves diarrhoea (D+) and the other, D− or atypical, does not.

D+ HUS is the classic form, accounting for 95% of cases of haemolytic uremic syndrome in children. This form of haemolytic uremic syndrome occurs predominantly in children and is preceded by a prodrome of diarrhoea, most commonly caused by an infection by shiga-toxin producing *Escherichia coli*. Specifically, *E. coli* serotype O157:H7 has been associated with more than 80% of infections leading to haemolytic uremic syndrome. The shiga-like toxin affects endothelial cells and initiates intravascular thrombogenesis. After entering the circulation via the gastrointestinal mucosa, the toxin preferentially localizes to the kidneys, inhibiting protein synthesis and eventually leading to cell necrosis or apoptosis. Endothelial cell damage subsequently potentiates renal microvascular thrombosis by promoting activation of the blood coagulation cascade. Platelet aggregation results in a consumptive thrombocytopenia. Microangiopathic haemolytic anaemia results from mechanical damage to red blood cells circulating through partially occluded microcirculation.

D – HUS accounts for the remaining 5% of cases of haemolytic uremic syndrome and its aetiology, age at onset, and clinical presentations are far more varied. Unlike D+ HUS, D – HUS is not preceded by an identifiable gastrointestinal infection. The pathogenesis of D – HUS has been the focus of current research and has, thus far, been associated with complement dysregulation in up to 50% of cases. Specifically, mutations in complement regulatory protein factor H, factor I, or factor B or autoantibodies against factor H have all been implicated. These mutations result in inability to suppress complement activation and for reasons that are not completely understood, the glomerular endothelium is particularly susceptible to these changes [23].

Case 54

Clinical presentations

- General features of anaemia
- Gastrointestinal tract bleeding
- Headache, hypertension, and confusion
- Fever

FBE RESULTS

CBC (Complete Blood Count) Parameters

Parameters	Results	Units
Hb	102	g/L
RCC	3.33	$\times 10^{12}\,L^{-1}$
PCV	0.33	
MCV	99	Fl
MCH	30.6	Pg
MCHC	309	g/L
RDW	23.5	%
Retics	N/A	%
WBC	15.1	$\times 10^{9}\,L^{-1}$
Plt count	15	$\times 10^{9}\,L^{-1}$

Differential White Cell Count

Cell Type	Percentages (%)	Absolute Count
Neutrophils	93	$14.04 \times 10^{9}\,L^{-1}$
Bands	3	$0.45 \times 10^{9}\,L^{-1}$
Lymphocytes	3	$0.45 \times 10^{9}\,L^{-1}$
Monocytes	1	$0.15 \times 10^{9}\,L^{-1}$
Eosinophils	0	$0 \times 10^{9}\,L^{-1}$
Basophils	0	$0 \times 10^{9}\,L^{-1}$

Differential White Cell Count—cont'd		
Cell Type	Percentages (%)	Absolute Count
Metamyelocytes	0	
Myelocytes	0	
Promyelocytes	0	
Blasts	0	
Prolymphocytes	0	
nRBC/100WBC	0	

RBC MORPHOLOGY

Moderate anisocytosis, moderate polychromasia, moderate hypochromic, mild fragments, and marked crenated cells.

WBC MORPHOLOGY

Mild leucocytosis, marked neutrophilia with mild toxic granulation.

PLATELET MORPHOLOGY

Marked thrombocytopenia with normal morphology.

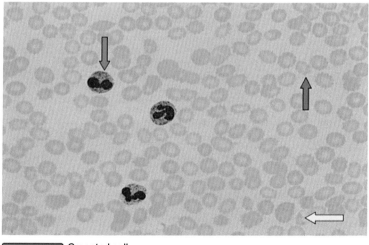

◼ Crenated cell
◼ Toxic granulation
▭ Fragment cell

Blood film image shows the most diagnostic features.

PROVISIONAL DIAGNOSIS

Based on clinical presentations, FBE results, and morphology, this case is highly suggestive of thrombotic thrombocytopenic purpura and further tests are suggested.

DIFFERENTIAL DIAGNOSES

- Idiopathic thrombocytopenic purpura
- Haemolytic uremic syndrome
- Disseminated intravascular coagulation
- Pregnancy-associated microangiopathy
- Transplant-associated thrombocytopenic purpura
- Haemolysis elevated liver enzyme and low platelets (HELLP)
- Drug-induced thrombocytopenia
- Malignancy
- Autoimmune disorders
- Sepsis/bacterial

FURTHER TESTS AND EXPECTED RESULTS

Confirmation of Provisional Diagnosis

Liver function tests:

- **LDH** → elevated
- **Unconjugated bilirubin** → increased
- **Haptoglobin** → low

Renal function tests:

- **Creatinine** → elevated
- **Urinalysis** → proteinuria and haematuria

  ADAMTS13 → decreased

Exclusion of Differential Diagnoses

Coagulation tests (PT, aPTT, D-dimer, and fibrinogen) → normal
Drug history → negative
Blood culture *E. coli* → negative
DAT → negative
Microbiology test for sepsis → negative

FINAL DIAGNOSIS

Thrombotic thrombocytopenic purpura (TTP) is the final diagnosis for this case based on further tests that confirm the provisional diagnosis and exclude all the differential diagnoses.

PATHOPHYSIOLOGY OF TTP

The TTP syndrome is characterized by microangiopathic haemolysis and platelet aggregation/hyaline thrombi whose formation is unrelated to coagulation system activity. Platelet microthrombi predominate; they form in the microcirculation (i.e. arterioles, capillaries) throughout the body causing partial occlusion of vessels. Organ ischaemia, thrombocytopenia, and erythrocyte fragmentation (i.e. schistocytes) occur. The thrombi partially occlude the vascular lumina with overlying proliferative endothelial cells. The endothelia of the kidneys, brain, heart, pancreas, spleen, and adrenal glands are particularly vulnerable to TTP. The liver, lungs, gastrointestinal tract, gall bladder, skeletal muscles, retina, pituitary gland, ovaries, uterus, and testes are also affected to a lesser extent. No inflammatory changes occur. The occlusion of the microthrombi affects many organs, and a myriad of symptoms are presented.

von Willebrand factor (vWF) is a large, adhesive glycoprotein that mediates thrombus formation at sites of vascular injury. vWF is synthesized in the endothelium and megakaryocytes, and it circulates in the plasma. Various sizes of multimers were noted, and the large form, ultralarge von Willebrand factor (ULVWF) multimers were secreted from the endothelium. These are the largest soluble protein found in human plasma and are considered the major pathogenic factor in TTP due to the platelet clumping in the microvasculature.

The ULVWF is the most active of the various-sized multimers and is found in platelets, endothelial cells, and subendothelium. This large vWF appeared to have a greater ability to adhere with platelets mediating a thrombus formation. The large vWF combine with platelets consumed from the arterioles and capillaries of organs in a high-shearing stress environment and cause endothelial injury leading to ischaemia. The red blood cells collide with the thrombi, and fragment leads to haemorrhage. As a result, the organ function is compromised. The agitated endothelial cells are the main source of ULVWF multimer secretion into the bloodstream where they bind to specific surface platelet receptors. The ULVWF multimers adhere to the damaged endothelium or exposed subendothelium, with the platelet receptor binding to the ULVWF. The sheer stress of fluid and platelet thrombi in the microcirculation does not enhance proteolysis of ULVWF but rather thrombi formation.

ADAMTS-13 is a metalloprotease consisting of multiple structural and functional domains and is the major regulator of the size of vWF in plasma. These domains may participate in the recognition and binding of

ADAMTS-13 to vWF. The ULVWF multimers are cleaved by ADAMTS-13 as they are secreted from endothelial cells.

Acquired TTP is associated with production of anti-ADAMTS13 antibodies inhibiting ADAMTS-13 activity. Congenital (familial) thrombotic thrombocytopenic purpura is associated with mutations of the vWF-cleaving protease ADAMTS-13 gene encoding, and ADAMTS-13 is inactivated or decreased. ADAMTS-13 is severely deficient in patients with both congenital TTP and acquired TTP [24].

Case 55

Clinical presentations

32-year-old pregnant woman has
- General features of anaemia
- Jaundice
- Hypertension
- Petechiae

FBE RESULTS

CBC (Complete Blood Count) Parameters

Parameters	Results	Units
Hb	95	g/L
RCC	3.46	$\times 10^{12} \, L^{-1}$
PCV	0.28	
MCV	81	Fl
MCH	27.5	Pg
MCHC	339	g/L
RDW	16.3	%
Retics	N/A	%
WBC	10.2	$\times 10^9 \, L^{-1}$
Plt count	45	$\times 10^9 \, L^{-1}$

Differential White Cell Count

Cell Type	Percentages (%)	Absolute Count
Neutrophils	78	$7.96 \times 10^9 \, L^{-1}$
Bands	10	$1.02 \times 10^9 \, L^{-1}$
Lymphocytes	9	$0.92 \times 10^9 \, L^{-1}$
Monocytes	3	$0.31 \times 10^9 \, L^{-1}$
Eosinophils	0	$0 \times 10^9 \, L^{-1}$

Continued

Differential White Cell Count—cont'd		
Cell Type	Percentages (%)	Absolute Count
Basophils	0	$0 \times 10^9 \, L^{-1}$
Metamyelocytes	0	
Myelocytes	0	
Promyelocytes	0	
Blasts	0	
Prolymphocytes	0	
nRBC/100WBC	0	

RBC MORPHOLOGY

Mild anisocytosis, mild polychromasia, mild spherocytes, and mild irregular contracted cells.

WBC MORPHOLOGY

Normal in number and morphology.

PLATELET MORPHOLOGY

Marked thrombocytopenia with normal morphology.

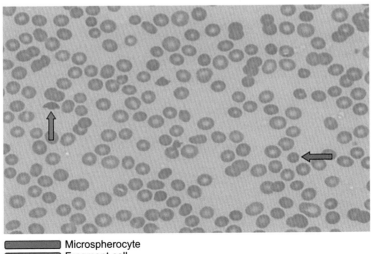

▭ Microspherocyte
▭ Fragment cell

Blood film image shows the most diagnostic features.

PROVISIONAL DIAGNOSIS

Based on clinical presentations, FBE results, and morphology, this case is highly suggestive of haemolysis, elevated liver enzyme, and low platelet (HELLP) and further tests are suggested.

DIFFERENTIAL DIAGNOSES

- Acute fatty liver of pregnancy
- Haemolytic uremic syndrome
- Thrombotic thrombocytopenic purpura

FURTHER TESTS AND EXPECTED RESULTS

Confirmation of Provisional Diagnosis

Liver function tests:

- **LDH** → elevated
- **Unconjugated bilirubin** → increased
- **Haptoglobin** → low
- **Serum transaminase** → increased

Exclusion of Differential Diagnoses

ADAMTS13 → normal
Coagulation tests (PT, aPTT, D-dimer, and fibrinogen) → normal
Blood culture *E. coli* → negative

FINAL DIAGNOSIS

Haemolysis elevated liver enzyme and low platelet (HELLP) is the final diagnosis for this case based on further tests that confirm the provisional diagnosis and exclude all the differential diagnoses.

Chapter 14

Haematological Infections

Case 56

Clinical presentations

- Lethargy
- Anorexia
- Headache
- Fever
- Lymphadenopathy and hepatosplenomegaly

FBE RESULTS

CBC (Complete Blood Count) Parameters		
Parameters	Results	Units
Hb	125	g/L
RCC	4.60	$\times 10^{12}\,L^{-1}$
PCV	0.39	
MCV	85	Fl
MCH	27.0	Pg
MCHC	321	g/L
RDW	13.8	%
Retics	N/A	%
WBC	9.2	$\times 10^{9}\,L^{-1}$
Plt count	206	$\times 10^{9}\,L^{-1}$

Haematology Case Studies with Blood Cell Morphology and Pathophysiology. http://dx.doi.org/10.1016/B978-0-12-811911-2.00014-3

Differential White Cell Count		
Cell type	Percentages (%)	Absolute count
Neutrophils	24	$2.21 \times 10^9 \, L^{-1}$
Bands	7	$0.64 \times 10^9 \, L^{-1}$
Lymphocytes	26	$2.39 \times 10^9 \, L^{-1}$
Monocytes	10	$0.92 \times 10^9 \, L^{-1}$
Eosinophils	0	$0 \times 10^9 \, L^{-1}$
Basophils	0	$0 \times 10^9 \, L^{-1}$
Metamyelocytes	0	
Myelocytes	0	
Promyelocytes	0	
Blasts	0	
Others	33	
nRBC/100WBC	0	

RBC MORPHOLOGY

Normocytic normochromic cells.

WBC MORPHOLOGY

Mild lymphocytosis and moderate atypical lymphocyte approximately 33%.

PLATELET MORPHOLOGY

Normal in number and morphology.

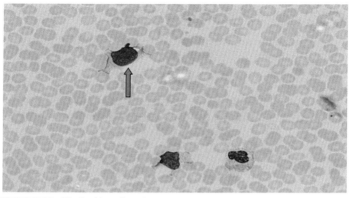

▭ Atypical lymphocyte

Blood film image shows the most diagnostic features.

PROVISIONAL DIAGNOSIS

Based on clinical presentations, FBE results, and morphology, this case is highly suggestive of infectious mononucleosis and further tests are suggested.

DIFFERENTIAL DIAGNOSES

Other viral infection
ALL (acute lymphoblastic leukaemia)
Non-Hodgkin's lymphoma
Bacterial infection
Protozoan infection
Systemic lupus erythematosus
Drug hypersensitivity

FURTHER TESTS AND EXPECTED RESULTS

Confirmation of Provisional Diagnosis

Paul–Bunnell (monospot) test → positive
Heterophile antibody → positive
C-reactive protein → high
ESR → high

Exclusion of Differential Diagnoses

Blood culture for bacteria → negative
Serology for toxoplasmosis → negative

Drug history → negative
Serology test for lupus → negative
Flow cytometry to exclude malignancies → negative

FINAL DIAGNOSIS

Infectious mononucleosis (IM) is the final diagnosis for this case based on further tests that confirm the provisional diagnosis and exclude all the differential diagnoses.

PATHOPHYSIOLOGY OF IM

The infectious mononucleosis caused by Epstein–Barr virus (EBV) infection. EBV is transmitted via intimate contact with body secretions, primarily oropharyngeal secretions. EBV infects the B-cells in the oropharyngeal epithelium. The organism may also be shed from the uterine cervix, implicating the role of genital transmission in some cases. On rare occasion, EBV is spread via blood transfusion. Circulating B-cells spread the infection throughout the entire reticular endothelial system (RES), i.e. liver, spleen, and peripheral lymph nodes. EBV infection of B-lymphocytes results in a humoral and cellular response to the virus. The humoral immune response directed against EBV structural proteins is the basis for the test used to diagnose EBV infectious mononucleosis. However, the T-lymphocyte response is essential in the control of EBV infection; natural killer (NK) cells and predominantly CD8$^+$ cytotoxic T-cells control proliferating B-lymphocytes infected with EBV. The T-lymphocyte cellular response is critical in determining the clinical expression of EBV viral infection. A rapid and efficient T-cell response results in control of the primary EBV infection and lifelong suppression of EBV. Ineffective T-cell response may result in excessive and uncontrolled B-cell proliferation, resulting in B-lymphocyte malignancies (e.g. B-cell lymphomas). The immune response to EBV infection is fever, which occurs because of cytokine release consequent to B-lymphocyte invasion by EBV. Lymphocytosis observed in the RES is caused by a proliferation of EBV-infected B-lymphocytes. Pharyngitis observed in EBV infectious mononucleosis is caused by the proliferation of EBV-infected B-lymphocytes in the lymphatic tissue of the oropharynx [25].

Case 57

Clinical presentations

- Fever
- Cold
- Cough
- Headache

FBE RESULTS

CBC (Complete Blood Count) Parameters

Parameters	Results	Units
Hb	108	g/L
RCC	3.37	$\times 10^{12}\ L^{-1}$
PCV	0.31	
MCV	92	Fl
MCH	32.0	Pg
MCHC	348	g/L
RDW	17.9	%
Retics	N/A	%
WBC	18.9	$\times 10^{9}\ L^{-1}$
Plt count	200	$\times 10^{9}\ L^{-1}$

Differential White Cell Count

Cell Type	Percentages (%)	Absolute Count
Neutrophils	27	$5.10 \times 10^{9}\ L^{-1}$
Bands	39	$7.37 \times 10^{9}\ L^{-1}$
Lymphocytes	8	$1.51 \times 10^{9}\ L^{-1}$
Monocytes	9	$1.70 \times 10^{9}\ L^{-1}$
Eosinophils	1	$0.19 \times 10^{9}\ L^{-1}$
Basophils	0	$0 \times 10^{9}\ L^{-1}$

Continued

Differential White Cell Count—cont'd		
Cell Type	Percentages (%)	Absolute Count
Metamyelocytes	8	
Myelocytes	8	
Promyelocytes	0	
Blasts	0	
Others	0	
nRBC/100WBC	0	

RBC MORPHOLOGY

Moderate anisocytosis, mild polychromasia, mild macrocytes, mild target cells, mild irregular contracted cells, and occasional nRBC.

WBC MORPHOLOGY

Mild leucocytosis, marked left shift, and marked toxic granulation.

PLATELET MORPHOLOGY

Normal in number and morphology.

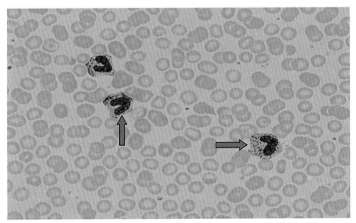

▬▬▬ Left shift (indicates that the neutrophils present in the blood are at a slightly earlier stage of maturation such as a band form)

▬▬▬ Toxic granulation (dark coarse granules found in the cytoplasm of the neutrophils)

Blood film image shows the most diagnostic features.

PROVISIONAL DIAGNOSIS

Based on clinical presentations, FBE results, and morphology, this case is highly suggestive of bacterial infection with marked left shift and toxic granulation and further tests are suggested.

DIFFERENTIAL DIAGNOSES

Other causes of infection such as viral, fungal, and parasitical infection. CML (chronic myelocytic leukaemia) and other myeloproliferative neoplasms.

FURTHER TESTS AND EXPECTED RESULTS

Confirmation of Provisional Diagnosis

Blood culture and other bacteriological tests → positive

Exclusion of Differential Diagnoses

Serology viral and fungal tests → negative exclude the viral and fungal infection
Cytogenetic
Philadelphia chromosome t(9:22) → negative exclude CML

FINAL DIAGNOSIS

Bacterial infection with marked left shift and toxic granulation is the final diagnosis for this case based on further tests that confirm the provisional diagnosis and exclude all the differential diagnoses.

Case 58

Clinical presentations

- Malaise, chills, and fever
- Thready pulse, headache, and nausea
- Anaemia and jaundice
- Diarrhoea
- Hepatosplenomegaly

FBE RESULTS

CBC (Complete Blood Count) Parameters

Parameters	Results	Units
Hb	122	g/L
RCC	4.55	$\times 10^{12}\,L^{-1}$
PCV	0.44	
MCV	97	Fl
MCH	27.0	Pg
MCHC	277	g/L
RDW	19.0	%
Retics	N/A	%
WBC	5.2	$\times 10^{9}\,L^{-1}$
Plt count	80	$\times 10^{9}\,L^{-1}$

Differential White Cell Count

Cell Type	Percentages (%)	Absolute Count
Neutrophils	45	$2.34 \times 10^{9}\,L^{-1}$
Bands	23	$1.20 \times 10^{9}\,L^{-1}$
Lymphocytes	20	$1.04 \times 10^{9}\,L^{-1}$
Monocytes	8	$0.42 \times 10^{9}\,L^{-1}$
Eosinophils	3	$0.16 \times 10^{9}\,L^{-1}$

Differential White Cell Count—cont'd

Cell Type	Percentages (%)	Absolute Count
Basophils	0	$0 \times 10^9 \, L^{-1}$
Metamyelocytes	1	
Myelocytes	0	
Promyelocytes	0	
Blasts	0	
Others	0	
nRBC/100WBC	0	

RBC MORPHOLOGY

Moderate anisocytosis, moderate parisitaemia approximately 3% showing trophozoites, merozoites, and schizont stages. Infected cells are enlarged and pale yellow/brown pigment, and occasional bubble rings.

WBC MORPHOLOGY

Mild left shift and mild toxic granulation.

PLATELET MORPHOLOGY

Marked thrombocytopenia with normal morphology.

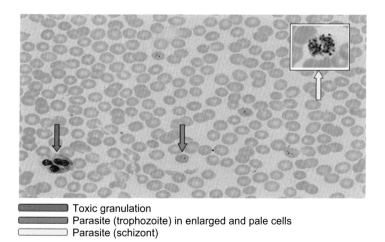

▬▬▬ Toxic granulation
▬▬▬ Parasite (trophozoite) in enlarged and pale cells
▭▭▭ Parasite (schizont)

Blood film image shows the most diagnostic features.

PROVISIONAL DIAGNOSIS

Based on clinical presentations, FBE results, and morphology, this case is highly suggestive of *Plasmodium vivax* (malarial) infection and further tests are suggested.

DIFFERENTIAL DIAGNOSES

Other malarial species and other parasitic infection
Infectious mononucleosis and other viral infection
Toxoplasmosis
Bacterial infection

FURTHER TESTS AND EXPECTED RESULTS

Confirmation of Provisional Diagnosis

Thick film → detect low parisitaemia
Polymerase chain reaction → detect *P. vivax*
Rapid diagnostic test such as immunochromatographic test (ICT) for P. vivax → positive
Parasite-specific LDH → distinguish between live and dead parasite

Exclusion of Differential Diagnoses

Rapid diagnostic test such as immunochromatographic test (ICT) for *Plasmodium falciparum* → negative
Blood culture for bacteria → negative
Paul–Bunnell (monospot) test → negative
Serology for toxoplasmosis → negative

FINAL DIAGNOSIS

P. vivax (malarial) infection is the final diagnosis for this case based on further tests that confirm the provisional diagnosis and exclude all the differential diagnoses.

Case 59

Clinical presentations

- Malaise, chills, and fever
- Thready pulse, headache, and nausea
- Anaemia and jaundice
- Diarrhoea
- Hepatosplenomegaly

FBE RESULTS

CBC (Complete Blood Count) Parameters

Parameters	Results	Units
Hb	124	g/L
RCC	4.43	$\times 10^{12}\,L^{-1}$
PCV	0.37	
MCV	86	Fl
MCH	28.0	Pg
MCHC	331	g/L
RDW	N/A	%
Retics	N/A	%
WBC	8.5	$\times 10^{9}\,L^{-1}$
Plt count	78	$\times 10^{9}\,L^{-1}$

Differential White Cell Count

Cell Type	Percentages (%)	Absolute Count
Neutrophils	58	$4.93 \times 10^{9}\,L^{-1}$
Bands	24	$2.04 \times 10^{9}\,L^{-1}$
Lymphocytes	11	$0.93 \times 10^{9}\,L^{-1}$
Monocytes	6	$0.51 \times 10^{9}\,L^{-1}$
Eosinophils	1	$0.08 \times 10^{9}\,L^{-1}$

Continued

Differential White Cell Count—cont'd		
Cell Type	Percentages (%)	Absolute Count
Basophils	0	$0 \times 10^9 \, L^{-1}$
Metamyelocytes	0	
Myelocytes	0	
Promyelocytes	0	
Blasts	0	
Others	0	
nRBC/100WBC	0	

RBC MORPHOLOGY

Normocytic normochromic cells, numerous ring forms with delicate umbrellas—some applique forms infected cells.

WBC MORPHOLOGY

Mild left shift with mild toxic vacuolation.

PLATELET MORPHOLOGY

Moderate thrombocytopenia with normal morphology.

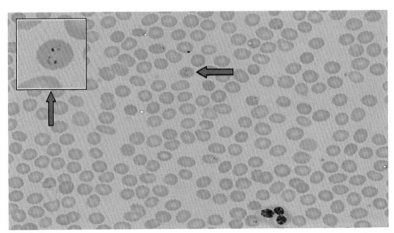

Ring forms of *Plasmodium falciparum* in normocytic normochromic cells

Blood film image shows the most diagnostic features.

PROVISIONAL DIAGNOSIS

Based on clinical presentations, FBE results, and morphology, this case is highly suggestive of *P. falciparum* (malarial) infection and further tests are suggested.

DIFFERENTIAL DIAGNOSES

Other malarial species and other parasitic infection
Infectious mononucleosis and other viral infection
Toxoplasmosis
Bacterial infection

FURTHER TESTS AND EXPECTED RESULTS

Confirmation of Provisional Diagnosis

Thick film → detect low parasitaemia
Polymerase chain reaction → detect *P. falciparum*
Rapid diagnostic test such as ICT → detect *P. falciparum*
Parasite-specific LDH → distinguish between live and dead parasite

Exclusion of Differential Diagnoses

Rapid diagnostic test such as ICT for *P. vivax* → negative
Blood culture for bacteria → negative
Paul–Bunnell (monospot) test → negative
Serology for toxoplasmosis → negative

FINAL DIAGNOSIS

P. falciparum (malarial) infection is the final diagnosis for this case based on further tests that confirm the provisional diagnosis and exclude all the differential diagnoses.

Case 60

Clinical presentations

- Malaise, chills, and fever
- Thready pulse, headache, and nausea
- Anaemia and jaundice
- Diarrhoea
- Hepatosplenomegaly

FBE RESULTS

CBC (Complete Blood Count) Parameters

Parameters	Results	Units
Hb	136	g/L
RCC	4.10	$\times 10^{12}\,L^{-1}$
PCV	0.39	
MCV	95	Fl
MCH	33.0	Pg
MCHC	349	g/L
RDW	22.0	%
Retics	N/A	%
WBC	6.1	$\times 10^{9}\,L^{-1}$
Plt count	180	$\times 10^{9}\,L^{-1}$

Differential White Cell Count

Cell Type	Percentages (%)	Absolute Count
Neutrophils	50	$3.05 \times 10^{9}\,L^{-1}$
Bands	6	$0.37 \times 10^{9}\,L^{-1}$
Lymphocytes	23	$1.40 \times 10^{9}\,L^{-1}$
Monocytes	21	$1.28 \times 10^{9}\,L^{-1}$
Eosinophils	0	$0 \times 10^{9}\,L^{-1}$

Differential White Cell Count—cont'd

Cell Type	Percentages (%)	Absolute Count
Basophils	0	$0 \times 10^9 \, L^{-1}$
Metamyelocytes	0	
Myelocytes	0	
Promyelocytes	0	
Blasts	0	
Others	0	
nRBC/100WBC	0	

RBC MORPHOLOGY

Marked anisocytosis, mild polychromasia, and mild malarial parasites.

WBC MORPHOLOGY

Mild reactive lymphocytes.

PLATELET MORPHOLOGY

Normal in number and morphology.

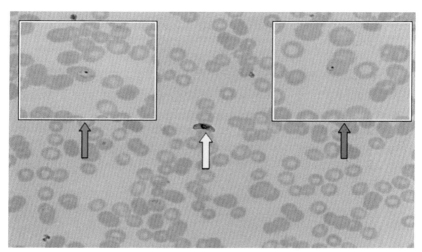

Delicate rings of *Plasmodium falciparum* in normocytic normochromic red blood cell

Plasmodium vivax in enlarged, pale red blood cell

Gametocyte (banana shape)

Blood film image shows the most diagnostic features.

PROVISIONAL DIAGNOSIS

Based on clinical presentations, FBE results, and morphology, this case is highly suggestive of *P. falciparum* combined with *P. vivax* (malarial) infection and further tests are suggested.

DIFFERENTIAL DIAGNOSES

Other malarial species and other parasitic infection
Infectious mononucleosis and other viral infection
Toxoplasmosis
Bacterial infection

FURTHER TESTS AND EXPECTED RESULTS

Confirmation of Provisional Diagnosis

Thick film → detect low parasitaemia
Polymerase chain reaction → detect *P. falciparum*
Rapid diagnostic test such as ICT → detect *P. falciparum* and *P. vivax*
Parasite-specific LDH → distinguish between live and dead parasite

Exclusion of Differential Diagnoses

Blood culture for bacteria → negative
Paul–Bunnell (monospot) test → negative
Serology for toxoplasmosis → negative

FINAL DIAGNOSIS

P. falciparum combined with *P. vivax* (malarial) infection is the final diagnosis for this case based on further tests that confirm the provisional diagnosis and exclude all the differential diagnoses.

Next two images are other less common malaria species

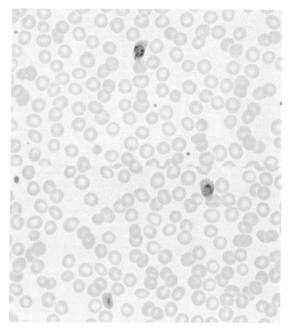

Plasmodium ovale

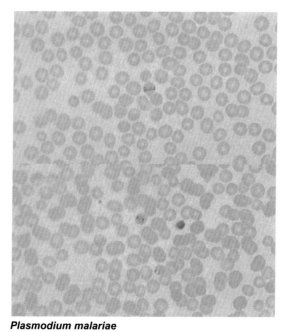

Plasmodium malariae

Please note unique shape of parasites in each case.

Features of different *Plasmodium* species

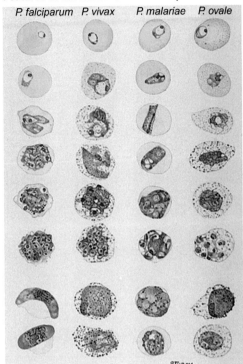

Malarial parasites from four species of *Plasmodium*. (From Diggs LW, Sturm D, Bell A: *Morphology of human blood cells*, ed 5, Abbott Park, Ill, 1985, Abbott Laboratories. Permission has been granted with approval of Abbott Laboratories, all rights reserved by Abbott Laboratories, Inc.)

PATHOPHYSIOLOGY OF MALARIAL INFECTION

Individuals with malaria typically acquired the infection in an endemic area following a mosquito bite. The risk of infection depends on the intensity of malaria transmission and the use of precautions such as bed nets, diethyl-meta-toluamide (DEET), and malaria prophylaxis. After a mosquito takes a blood meal, the malarial sporozoites enter hepatocytes (liver phase) within minutes and then emerge in the bloodstream after a few weeks. These merozoites rapidly enter erythrocytes and develop into trophozoites and then into schizonts over a period of days inside erythrocytes during the erythrocytic phase of the life cycle. Rupture of infected erythrocytes containing the schizont results in fever and merozoite release. The merozoites enter new red cells, and the process is repeated, resulting in a logarithmic increase in parasite burden. The outcome of infection depends on host immunity. Individuals with immunity can spontaneously clear the parasites. In those without immunity, the parasites continue to expand the infection. *P. falciparum* infection can result in death. A small percentage of parasites become gametocytes, which undergo sexual reproduction when taken up by the mosquito. These can develop into infective sporozoites, which continue the transmission cycle after a blood meal in a new host. Each *Plasmodium* species has a specific incubation period. Reviews of travellers returning from endemic areas have reported that *P. falciparum* infection typically develops within one month of exposure, thereby establishing the basis for continuing antimalarial prophylaxis for 4 weeks upon return from an endemic area. This should be emphasized to the patient to enhance posttravel compliance. Rarely, *P. falciparum* causes initial infection up to a year later. *P. vivax* and *P. ovale* may emerge weeks to months after the initial infection. In addition, *P. vivax* and *P. ovale* have a hypnozoite form during which the parasite can linger in the liver for months before emerging and inducing recurrence after the initial infection. In addition to treating the organism in infected blood, treating the hypnozoite form with a second agent (primaquine) is critical to prevent relapse from this latent liver stage. *P. falciparum* infection typically causes severe malaria. This species is more virulent because it may create high levels of parasitaemia and sequestration that contribute to end-organ damage. Sequestration is a specific property of this species. As it develops through the 48-hour life cycle, it demonstrates adherence properties, which result in the sequestration of the parasite in small postcapillary vessels. For this reason, only early forms are observed in the peripheral blood, before the sequestration property develops; this is an important diagnostic clue that the patient is infected with *P. falciparum*. Sequestration of parasites may contribute to mental status changes and coma, observed exclusively in *P. falciparum* infection. In addition, cytokines and a high burden of parasites contribute to end-organ disease. End-organ disease may develop rapidly in patients with *P. falciparum* infection, and it specifically involves the central nervous system (CNS), lungs, and kidneys. Other manifestations of *P. falciparum* infection include hypoglycaemia, lactic acidosis, severe anaemia, and multiorgan dysfunction due to hypoxia [26].

CONCLUSION

The blood film has become an important tool even in this present era of molecular analysis. It is pertinent for doctors to request blood smear from haematology laboratory when they are clinically indicated. The laboratory technicians and scientists need to prepare and carefully examine a blood film whenever the results of a full blood count are abnormal. To avoid error, high-tech and sophisticated investigations of hematologic disorders need to be interpreted in view of both peripheral blood film morphology as well as the clinical context. Moreover, virtual imaging is the future of haematology laboratories with molecular diagnosis.

Haematology and Chemistry Reference Intervals

Parameters	Normal Ranges	Units
RBC count		
Male	4.5–6.5	$\times 10^{12} \, L^{-1}$
Female	3.8–5.8	$\times 10^{12} \, L^{-1}$
Neonate (full term)	4.0–6.0	$\times 10^{12} \, L^{-1}$
Haemoglobin		
Male	130–180	g/L
Female	115–165	g/L
Neonate (full term)	135–195	g/L
Haematocrit/packed cell volume (PCV)		
Male	0.40–0.54	
Female	0.37–0.47	
Neonate (1 day)	0.44–0.64	
Mean cell volume (MCV)		
Adults	80–100	Fl
Neonates	88–125	Fl
Mean cell haemoglobin (MCH)		
Adults	27.0–32	pg
Mean cell haemoglobin concentration (MCHC)	300–350	g/L
Erythrocyte sedimentation rate (ESR)		
Male 17–50 y/o →	1–10	mm/h
More than 50 y/o →	2–14	mm/h
Female 17–50 y/o →	3–19	mm/h
More than 50 y/o →	5–20	mm/h
Children →	2–15	mm/h
Reticulocyte count		
Adults	0.2–2.0	%
Neonates	2.0–6.0	%
Red blood cell distribution width (RDW)	11.0–15.0	%
Platelet count	150–400	$\times 10^{9} \, L^{-1}$

Haematology and Chemistry Reference Intervals—cont'd

Parameters	Normal Ranges	Units
Leukocyte count		
Adults	4.0–11.0	$\times 10^9 \, L^{-1}$
Neonate (full term)	6.0–22.0	$\times 10^9 \, L^{-1}$
Differential white cell count (absolute number)		
Adults		
Neutrophils	2.0–7.0	$\times 10^9 \, L^{-1}$
Bands	0.0–0.5	$\times 10^9 \, L^{-1}$
Lymphocytes	1.5–4.0	$\times 10^9 \, L^{-1}$
Monocytes	0.2–0.8	$\times 10^9 \, L^{-1}$
Eosinophils	0.04–0.4	$\times 10^9 \, L^{-1}$
Basophils	≤ 0.1	$\times 10^9 \, L^{-1}$
Iron (Fe)	10–30	µmol/L
Ferritin		
Male	30–300	µg/L
Female	15–200	µg/L
Transferrin	1.7–3.0	g/L
Iron binding capacity	45–80	µmol/L
Transferrin saturation	15–45	%
Vitamin B12	120–680	pmol/L
Folate	7–40	nmol/L
Total protein		
Adults	62–80	g/L
Neonate	40–75	g/L
Direct bilirubin		
Adults	<7	µmol/L
Neonate	<7	µmol/L
Total bilirubin		
Adults	<20	µmol/L
Neonate	<20	µmol/L
Urea (BUN)		
Adults	3.0–8.0	mmol/L

Continued

Haematology and Chemistry Reference Intervals—cont'd

Parameters	Normal Ranges	Units
Neonate	1.0–4.0	mmol/L
Creatinine		
Adult male	0.06–0.12	mmol/L
Adult female	0.05–0.11	mmol/L
Neonate	0.04–0.08	mmol/L
Alanine transaminase (ALT)		
Adults	<35	U/L
Neonate	<50	U/L
Alkaline phosphatase (ALP)		
Adults	25–100	U/L
Neonate	50–300	U/L
Lactate dehydrgenase (LDH)	110–230	U/L
Prothrombin time (PT)	11.0–15.0	s
Activated partial prothrombin time (APTT)	25.0–40.0	s
Thrombin clotting time (TCT)	14.0–18.0	s
Fibrinogen (FIB)	1.5–4.0	g/L
D-dimers	Less than 0.20	mg/L

Data from Royal College of Pathologists of Australia (RCPA) Manual Version 4.0, 2004.

References

[1] Bain BJ. Diagnosis from the blood smear. N Engl J Med 2005;353:498–507.

[2] Rodak B, Fritsma GA, Doig K. Hematology: clinical principles & applications. 3rd ed. St. Louis, MO: Saunders Elsevier; 2007.

[3] Bain BJ. Blood cells: a practical guide. 4th ed. Chichester, UK: Blackwell Publishing Ltd; 2006.

[4] Braunstein EM. The Merck Manuals Online Medical Library, Anemias Caused by Deficient Erythropoiesis: Iron Deficiency Anemia, Anemia of Chronic Blood Loss; Chlorosis. White-house Station, NJ: Merck Sharp & Dohme Co. Available from: http://www.msdmanuals.com/en-au/professional/hematology-and-oncology/anemias-caused-bydeficient-erythropoiesis/iron-deficiency-anemia#resources [updated 26 November 2016; cited 5 May 2017].

[5] Cheerva AC, Bleibel SA, Jones-Crawford JL, Kutlar A, Leonard RJ, Raj AB. Alpha Thalassemia. emedicine. Available from: http://emedicine.medscape.com/article/955496-clinical [updated 27 February 2017; cited 5 May 2017].

[6] Advani P, Talavera F, Conrad ME. Beta Thalassemia. emedicine. Available from: http://emedicine.medscape.com/article/206490-overview [updated 8 November 2016; cited 5 May 2017].

[7] Mir MA, Logue GL. Sideroblastic Anemias. emedicine. Available from: http://emedicine.medscape.com/article/1389794-overview [updated 18 November 2015; cited 5 May 2017].

[8] Maakaron JE, Taher AT. Sickle Cell Anemia. emedicine. Available from: http://emedicine.medscape.com/article/205926-overview [updated 3 October 2016; cited 5 May 2017].

[9] Lerma EV, Stein R. Anemia of Chronic Disease and Renal Failure. emedicine. Available from: http://emedicine.medscape.com/article/1389854-overview [updated 15 February 2015; cited 5 May 2017].

[10] Herrin VE. Macrocytosis. emedicine. Available from: http://emedicine.medscape.com/article/203858-overview [updated 24 September 2016; cited 5 May 2017].

[11] Schick P. Megaloblastic Anemia. emedicine. Available from: http://emedicine.medscape.com/article/204066-overview [updated 6 January 2017; cited 5 May 2017].

[12] Hoffbrand AV, Moss PAH, Pettit JE. Essential haematology. 5th ed. Chichester, UK: Blackwell Publishing Ltd; 2006.

[13] Seiter K. Acute Myelogenous Leukemia. emedicine. Available from: http://emedicine.medscape.com/article/197802-overview [updated 9 May 2017; cited 15 May 2017].

[14] Seiter K. Acute Lymphoblastic Leukemia. emedicine. Available from: http://emedicine.medscape.com/article/207631-overview [updated 22 March 2017; cited 5 May 2017].

[15] Besa EC, Krishnan K. Chronic Myelogenous Leukemia. emedicine. Available from: http://emedicine.medscape.com/article/199425-overview [updated 11 March 2017; cited 5 May 2017].

[16] Nagalla S, Besa EC. Polycythemia Vera. emedicine. Available from: http://emedicine.medscape.com/article/205114-overview [updated 2 December 2016; cited 6 May 2017].

[17] Lal A. Primary Myelofibrosis. emedicine. Available from: http://emedicine.medscape.com/article/197954-overview [updated 30 January 2017; cited 6 May 2017].

[18] Mir MA, Liu D, Patel SC, Rasool HJ. Chronic Lymphocytic Leukemia. emedicine. Available from: http://emedicine.medscape.com/article/199313-overview [updated 3 February 2017; cited 5 May 2017].

[19] Besa EC. Hairy Cell Leukemia. emedicine. Available from: http://emedicine.medscape.com/article/200580-overview [updated 28 April 2015; cited 5 May 2017].

[20] Abbasi MR, Sparano JA. Mantle Cell Lymphoma. emedicine. Available from: http://emedicine.medscape.com/article/203085-overview [updated 31 July 2015; cited 6 May 2017].

[21] Shah D, Seiter K. Multiple Myeloma. emedicine. Available from: http://emedicine.medscape.com/article/204369-overview [updated 8 December 2016; cited 5 May 2017].

[22] Levi MM, Schmaier AH. Disseminated Intravascular Coagulation. emedicine. Available from: http://emedicine.medscape.com/article/199627-overview [updated 29 September 2016; cited 4 May 2017].

[23] Tan AJ, Silverberg MA. Hemolytic Uremic Syndrome. emedicine. Available from: http://emedicine.medscape.com/article/779218-overview [updated 21 January 2015; cited 4 May 2017].

[24] Wun T. Thrombotic Thrombocytopenic Purpura. emedicine. Available from: http://emedicine.medscape.com/article/206598-overview [updated 20 June 2016; cited 5 May 2017].

[25] Cunha BA. Infectious Mononucleosis. emedicine. Available from: http://emedicine.medscape.com/article/222040-overview [updated 19 October 2016; cited 5 May 2017].

[26] Herchline TE, Simon RQ. Malaria. emedicine. Available from: http://emedicine.medscape.com/article/221134-overview [updated 26 October 2016; cited 6 May 2017].

Index

Note: Page numbers followed by *t* indicate tables.

Printed in the United States
By Bookmasters